U0036771

作者序

✤ 大數據如何入門？

大數據的時代，各行各業都在試著往大數據前進，將大數據導入企業，那最重要的是什麼呢？模型嗎？筆者認為是資料，沒有資料，一切都是空談，因此基礎工程、基礎建設尤為重要，那要如何完善基礎工程呢？這點就是本書談的核心，資料工程。

✤ 大數據產品

本書從最基本的資料收集、爬蟲開始，到資料庫、RESTful API、分散式，最後走到視覺化，完整的呈現，大數據產品的發展過程。筆者除了做開源資料，為大數據盡一份心力外，也希望將過程，寫成一本書，分享給大眾，希望提供一個入門磚，且不只是入門，本書所用到的技術，已經到 Senior 等級。另外，本書不單純以技術為主軸，而是引入真實案例、真實情境，讓讀者更能體會到，為什麼要使用這些技術，是為了解決什麼樣的問題。

✤ Side Project

工程師、分析師，除了專研技術以外，發展個人的 Side Project，也漸漸成為主流，筆者的專案，FinMind，也是從 Side Project 慢慢發展起來，目前在 GitHub 得到 1,900 stars，算是得到一些認可。開發 Side Project 有什麼好處呢？除了對履歷上加分外，持續開發，接觸的層面會更多更廣，對於架構設計、產品開發上，更加有經驗。且不只是接觸技術，而是有真實的案例，可以證明，你會這項技術，並應用在真實場景。另外，工程師最大的優勢是，有能力做出產品，甚至是分析能力，都有可能幫助讀者，創造出個人產品、品牌，筆者希望能借助本書，讓更多讀者進入這個領域，更多讀者開始建立個人 Project、產品、品牌。

最後，感謝深智數位出版社，邀請筆者撰寫這本書，讓筆者有機會，分享本書內容、產品開發經驗、大數據技術，給各位讀者。

作者簡介

林子軒，Sam，目前任職 17 LIVE 資料工程師。擅長資料工程、資料分析，希望對 Python 社群、大數據領域，提供一份心力。

經歷

- 17 LIVE 資料工程師。
- 曾任職永豐金證券，軟體工程師。
- 曾於 Open UP Summit 2019，擔任 Speaker。
- 曾任職 Tripresso，資料工程師。
- 東華研究所，應用數學碩士。

FinMind

- https://github.com/FinMind/FinMind
- https://finmindtrade.com/

個人 GitHub

- https://github.com/linsamtw

Email

- samlin266118@gmail.com

如對本書有疑問，歡迎寄信到以上信箱。

筆者除了本書內容外，GitHub 上也有寫文章介紹，兩篇 Kaggle 競賽經驗，分別是生產線分析、庫存需求預設，如對以上有興趣，也可寄信到以上信箱。

目錄

06 資料提供—RESTful API 設計

07　容器管理工具 Docker

第 2 篇　產品迭代 -- 測試運維

08 自動化測試

09 CICD 持續性整合、部屬

第 5 篇　排程管理工具

12　排程管理工具 - Apache Airflow

13　Redis 介紹

第 6 篇　監控系統

14 監控工具介紹

15 結論

第 1 篇
資料工程 ETL

Chapter

01
本書介紹

本書主要介紹在 GitHub 上，獲得 1,900 stars 的開源專案 --- FinMind 的架構，獨家解析內部使用的技術。該專案以大數據、資料工程為起點，開發出許多數據服務，包含 Python Package、API、視覺化分析、網站等，呈現產品從 0 到 1 的完整過程，看完此書，你將有能力打造自己的產品、作品，可以作為一個進入大數據領域的入門磚。

讀完本書後，你將完成一個以 Data 為主的 Side Project，包含資料收集、資料庫、API、視覺化呈現，一整套從 0 到 1 的作品，細節如下。

❑ 資料收集

1. Python 環境、套件管理工具 pipenv。
2. 證交所、櫃買中心、期交所爬蟲。
3. 使用 RabbitMQ、Flower、Celery 分散式爬蟲。

❑ 資料庫

4. 使用 Docker 架設 MySQL 資料庫。

❑ Api

5. 使用 FastAPI 撰寫 RESTful Api，並用 Docker 包裝成 Image。
6. 使用 No-IP 申請免費網址、使用 Let's Encrypt 免費的 SSL 憑證。

❑ 視覺化

7. Redash 業界視覺化工具。

❑ 排程管理工具

8. 分散式 Airflow。

❑ 一站式監控系統

9. Prometheus、Grafana。

❑ 其它

10. Linode 雲端。
11. Docker、Docker-Compose 與 Docker Swarm 現代化集群架構。
12. 爬蟲、API Unit Test 單元測試。

13. 使用 GitLab-CI、Gitlab-Runner 建立 CICD 自動化流程。

對於初學者來說，擁有以上技能，對於尋找初階 Data 相關工作，是很容易的（讀者可以搜尋 data 相關職缺）。而對於已經是 Junior，並想往 Senior 發展的讀者，Docker、Docker Swarm、Gitlab-CI、CICD、Unit Test 等工具，對讀者來說非常有幫助，可以協助完成高品質的程式，並藉由持續開發 Side Project，累積相關經驗。

筆者初期是資料科學背景，深深體會沒有資料的問題，因此慢慢走向資料工程，自動化收集資料，中期希望將分析結果呈現給更多人觀看，單純視覺化不夠，又跑去寫 Web，最後為了達到產品穩定性，做了許多維運的工，書中將會介紹每一步的過程。

FinMind 介紹

近年大數據技術蓬勃發展，許多人都嚮往成為資料科學家，而在資料科學的道路上，如果沒有好的資料工程作為基礎，沒有好的資料源，就像一台很強的火箭，卻沒有燃料，也是空談。

因此，FinMind 收集各種股市資料，並提供一般大眾直接拿取資料，可以寫程式下載資料，也可直接在網站上下載，解決了火箭燃料的問題，而本書中，會解析 FinMind 使用的資料工程技術。

該開源專案主要是使用分散式架構、爬蟲，進行收集資料、資料清洗、存入資料庫，並開發 API 供一般大眾使用。書中除了介紹最基本的爬蟲外，分散式、資料庫架設、API 開發、Loading Balance 等，都會一併介紹，並使用 Docker 進行多台機器的控管，在寫書的當下，FinMind 使用 9 台雲端機器，而如何控管這些機器，也會是書中介紹的重點之一。

產品架構圖

下圖是 FinMind 產品開發中，所使用到的技術，從最初的資料收集，到最後的 API、資料分析、視覺化、網站呈現、金流串接。

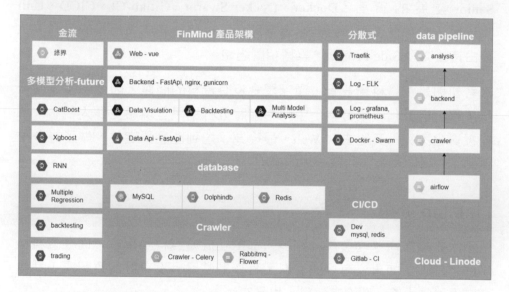

註 1：一般程式開發工具，常見的都是以黑底為主，而由於書本印刷原因，將以白底呈現，這點請讀者特別留意。

註 2：一般程式的排版，長度會以 80 為主，本書因為排版問題，長度會以 40 為主。

開發環境

2.1 開發環境重要性

首先,先介紹開發環境設定,有適合的作業系統、適合的開發工具,會讓你事半功倍,以下將介紹業界常用的開發工具。

2.2 Linux 作業系統

建議開發環境使用 Linux,市面上大多數服務、網站,都使用 Linux 系統,且雲端大多都提供 Linux 環境,因此直接在 Linux 下做開發,這樣對於產品部屬上,不容易遇到環境問題。如果只有 Windows 環境的話,下一個章節,會介紹如何在 Windows 環境下開發。

本書建議 Linux 的讀者使用 Ubuntu 18.04，這算是最知名的 Linux 作業系統之一，且相對來說，發生問題時，網路上的資源相對較多。

2.3 Windows 作業系統

過去如果要在 Windows 上使用 Linux 環境開發的話，大多會借助 VM 等虛擬機器，但多數虛擬機效能都不好。近年微軟開發出 WSL（Windows Subsystem for Linux）這套工具，可以讓開發者，在 Windows 上直接使用 Linux 環境，以下將介紹如何在 Windows 使用 WSL。

1. 準備 win10 專業版。

2. 開啟 [控制台]，點選 [程式和功能]。

3. 點選 [開啟或關閉 Windows 功能]。

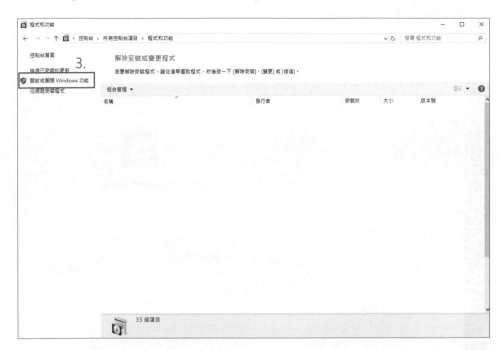

4. 勾選，[適用於 Linux 的 Windows 子系統]。

以上，已經啟用 Windows 中 Linux 子系統功能，接下來將安裝其中一個 Linux 系統 Ubuntu。

5. 進入 Microsoft Store

6. 搜尋 Ubuntu

7. 安裝 Ubuntu

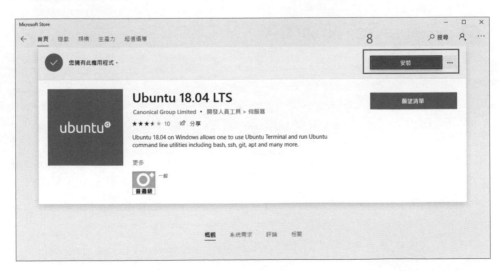

8. 安裝完後，點啟動。

9. 啟動完後，就會顯示此畫面，就有 Linux 環境啦！

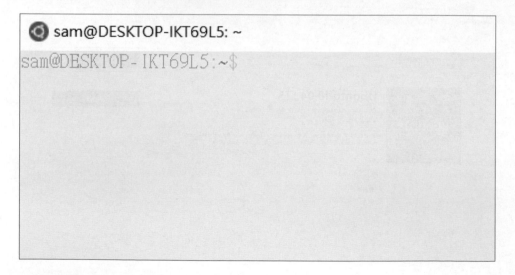

2.4　Mac 作業系統

如果是 Mac 的讀者，由於 Mac 操作上與 Linux 相近，因此不需要做額外設定，可直接進入 2.5 章節。

2.5　Python 開發工具 VS Code

好的 IDE 在開發上會事半功倍，本書使用 VS Code 這套 IDE 開發工具，因為它功能豐富，請讀者到以下連結下載 VS Code，本書使用的版本是1.41.1。

https://code.visualstudio.com/updates/v1_41

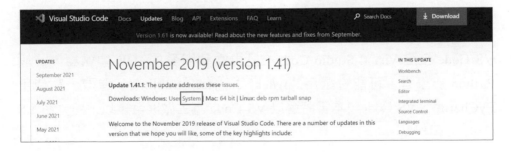

Windows 的讀者，請選擇 System，Mac 直接選擇 64 bit，Ubuntu 讀者選擇 deb 後安裝。

安裝完成後,如果是 Ubuntu、Mac 的讀者,就可以直接在 VS Code 的 TERMINAL 使用原生 Mac、Ubuntu 指令,如果是 Windows 環境,VS Code 需要做以下設定,在右下角,點選,**Select Default Shell**,並選擇 bash,之後點選垃圾桶,重新打開 TERMINAL,就會是 bash 了。

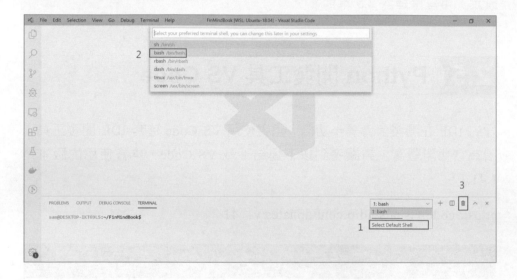

VS Code 全名 Visual Studio Code,是由微軟開發的 IDE 編輯器,一般 Python 初學者,可能會選用 Spyder、Jupyter、Sublime,進階一點使用 PyCharm、VS Code,本書將會以 VS Code 為主做開發,以下將介紹 VS Code 幾個優點。

❏ 輕量

相對於 PyCharm 來說,VS Code 相對輕量,開啟速度快,可以加快進入開發環境中,本書會開發多個專案,需要開啟多個 IDE,因此效能也是考量之一。

❏ 插件

VS Code 有豐富的插件，可額外下載一些工具，對你的 VS Code 進行
擴充，例如你要跑測試、觀看 Git Graph、甚至使用 Tabnine 這項 AI
Autocomplete 自動補齊的工具。

❏ 測試

在開發過程中,一定會經歷需要跑測試的環節,初學者可能不清楚什麼是測試,這在接下來的章節中會提到。而 Spyder、Jupyter、Sublime 都沒有測試的功能。

Chapter

03

Docker

3.1 為什麼先介紹 Docker ？

在開發 Python 專案時，初期可能只會停留在寫 code 階段，到了中期時，你會需要許多工具來幫助你。

以資料工程為例，初期收集資料，需架設資料庫來做儲存，如 MySQL，進階一點做分散式爬蟲，如 RabbitMQ、Flower，而安裝這些工具，需要一系列繁瑣的過程，如果你在 Linux，那安裝過程稍微好一點，如果在 Windows，更加繁瑣，借助 Docker，可以一行指令架好這些工具，在 4.3 將會介紹如何一行指令安裝資料庫。

3.2 什麼是 Docker？

在使用 Docker 之前，本書先介紹什麼是 Docker。

Docker 類似於虛擬機的概念，幫你分離出獨立環境，在產品開發上，可以在本地端模擬實際部屬的情境，當你在本地端開發完成後，可以無痛部屬到伺服器 / 雲端，而初期會遇到以下 Docker 相關名詞，本書將一一介紹。

Image

映像檔，如果有用過 VM 虛擬機，那應該很熟悉這名詞，Docker 一切的 Container，都是根據 Image 所建立的，如果要建立資料庫等服務，只需要在 Docker Hub 上，下載由其他人建好的 Image，就能直接啟動相關服務，Image 可以看成 Docker 最基本的元素，小到單一服務，大到整個架構，都是由最基礎的 Image 堆疊出來的。

Docker Hub

由 Docker 官方所建立的 Image 圖書館、社區，以下引述官方網站介紹：

> *Docker Hub is the world's largest library and community for container images.*

裡面包含了 Ubuntu、MySQL、Mongodb、Python、PostgresSQL、Redis 等各種服務的 Image，在過去，安裝以上服務，都需要不同的安裝方式，到了 Docker 的世界，你只需要學習 Docker 指令，就可以統一應用在架設這些服務上。

Container

容器，是根據 Image 所建立出對應的服務，例如使用 MySQL 的 Image，就可以建立出 MySQL 的 Container，這樣你就有 MySQL 資料庫可以使用了。

Volume

在使用 Docker 啟動資料庫的情境下，希望資料能永久保存，不會因為 Container 突然關閉，造成資料遺失，這時就需要借助 Volume，將 Container 內部指令的空間，與外部實體機做連接，這樣當 Container 需要做更新時，關閉重啟，資料庫的資料依然會存在。

讀者可能會想問，Docker 與 VM 有什麼差別？ VM 比較偏向作業系統層面，而 Docker 偏向應用服務層面。

對於初學者來說，Docker 最大的好處，是簡單的幫你安裝好各種服務，以分散式爬蟲為例，你會需要資料庫 MySQL、分散式工具 RabbitMQ，這些 Docker 都能非常簡易的協助你部屬，而 VM 做不到這點。

對於進階者來說，你可能會接觸到 CICD、Docker Swarm、K8s 等架構，這些都是以 Docker 為基礎下去進行，如測試、上線部屬等。以現代架構下，一定會接觸到分散式，Web、API、Backend 等服務，都是分散在多台機器上，借助 Docker，你可以一鍵更新所有機器上的服務，這也是 VM 做不到的。

3.3　安裝 Docker

以下將分別介紹，Ubuntu 跟 Windows 環境下，Docker 的安裝方法。

Ubuntu

打開 terminal，以下指令皆在 terminal 視窗下執行（也可在 VS Code 下的 terminal 執行）。

安裝

```
sudo apt-get update
sudo apt-get install -y docker.io
```

將使用者加入 Docker group

```
sudo usermod -aG docker user_name
```

這時需要先登出再重新登入，或是關閉這個 terminal，重新開新的一個，之後測試 Docker。

```
docker ps
```

結果如下圖，代表安裝成功。

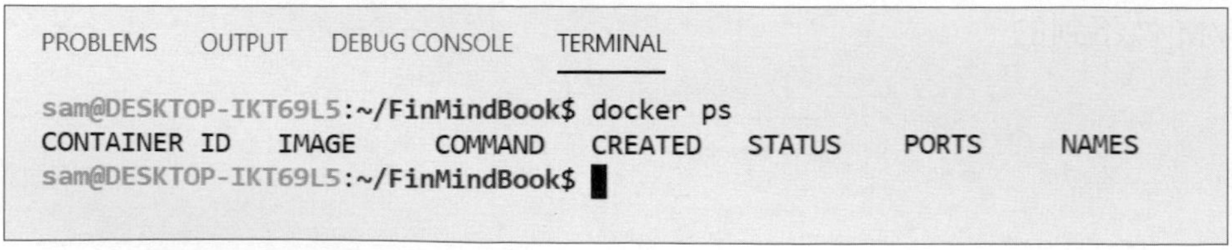

```
PROBLEMS    OUTPUT    DEBUG CONSOLE    TERMINAL

sam@DESKTOP-IKT69L5:~/FinMindBook$ docker ps
CONTAINER ID    IMAGE    COMMAND    CREATED    STATUS    PORTS    NAMES
sam@DESKTOP-IKT69L5:~/FinMindBook$ █
```

以上完成了 Linux 上 Docker 的安裝。

Windows

到以下連結下載 Docker：

https://docs.docker.com/docker-for-windows/install/

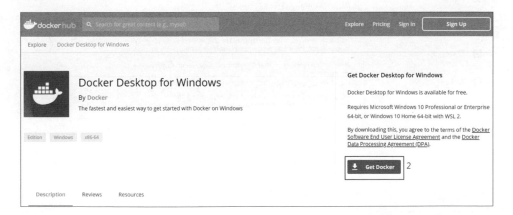

安裝完後，需額外設定 WSL。

目前在 VS Code terminal 視窗下,執行以下指令:

```
docker ps
```

會出現以下失敗訊息:

```
cannot connect to the docker daemon at unix ///var/run/docker.sock. is
the docker daemon running
```

需要先打開 Docker 設定,勾選以下選項:

Expose daemon on tcp://localhost:2375 without tls

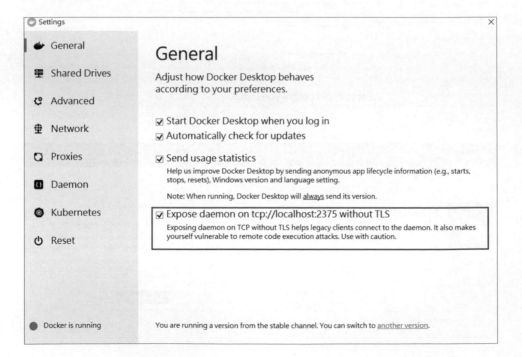

並將以下指令添加到 ~/.bashrc

```
export DOCKER_HOST=127.0.0.1:2375
```

開一個新的 terminal 視窗，並執行：

```
docker ps
```

如果結果與下圖相同，代表安裝成功：

```
PROBLEMS    OUTPUT    DEBUG CONSOLE    TERMINAL

sam@DESKTOP-IKT69L5:~/FinMindBook$ docker ps
CONTAINER ID    IMAGE      COMMAND    CREATED    STATUS    PORTS    NAMES
sam@DESKTOP-IKT69L5:~/FinMindBook$ █
```

如果是新版 Windows 的讀者，可能會遇到 WSL2 的問題，本書用的是 WSL 1，如果是 WSL 2 的讀者，請參考以下連結，否則在 docker 的部分，會遇到問題。

https://docs.microsoft.com/zh-tw/windows/wsl/tutorials/wsl-containers

3.4 安裝 Docker-Compose

根據官方文件 https://docs.docker.com/compose/install/，Linux 部分，使用以下指令安裝：

```
sudo curl -L "https://github.com/docker/compose/releases/
download/1.29.1/
docker-compose-$(uname -s)-$(uname -m)" -o /usr/local/bin/docker-compose

sudo chmod +x /usr/local/bin/docker-compose
```

Windows 部分，只要安裝的 Docker Desktop 即可。

有 Docker 之後，本書用到的服務，全部都是一行指令安裝完成。接下來將介紹 ETL 資料工程幾個步驟，資料收集（爬蟲）、資料清理、存入資料庫，但是在這之前，先介紹雲端，以便後續實際應用的案例分享。

Chapter

04

雲端

4.1 為什麼要用雲端？

在 Chapter 5，將介紹分散式爬蟲，而分散式當然不可能只用一台電腦完成，至少需要 2 台電腦做分散，一般人不會有 2 台電腦，這時雲端是個好選擇，本章節介紹一個非常便宜的雲端解決方案，Linode，最低一個月一台雲端機器只要 5 美金，24 小時不斷電、不斷網、不用維護，以自己架設伺服器來說，光網路費就超過 5 美金了。

https://www.linode.com/pricing/

以下擷取 Linode 部分內容。

RAM	CPU	Storage	Transfer	Network In	Network Out	Price	
1 GB	1 Core	25 GB SSD	1 TB	40 Gbps	1000 Mbps	$5 / mo	($.0075 / hr)

這時讀者不用擔心，是不是馬上要花錢了呢？ GCP、AWS 都有提供免費額度，同樣的，Linode 也有，筆者這裡準備一個 Linode 網址，點進去註冊，就能獲得 100 美金的額度。

https://www.linode.com/lp/free-credit-short/

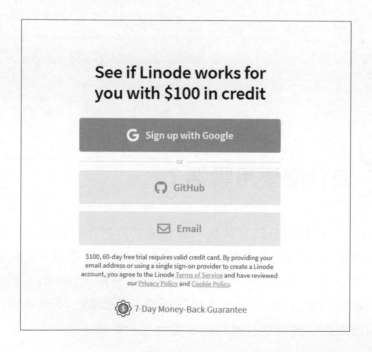

進入連結後，會看到上圖，使用這連結註冊，馬上會獲得 100 美金的額度，足夠讀者同時開 3 台機器（一台 5 美金，三台 15 美金），運行半年以上的時間。

另外的重點是，Linode 每台都提供固定 **IP**，2 台機器就有 2 個 IP，搭配 Python 的 sleep 函數，不用擔心爬蟲被 ban IP 了，也不用擔心用 sleep 效能差，多開幾台就解決了。

下一章節將介紹分散式爬蟲，初期可以開兩台機器嘗試，300 台幣 / 月（10 美金 / 月），Linode 也可以根據使用的小時（0.0075/hr）收費，做完本書範例後，不想使用，也可以刪除，就不用付到一個月的錢了。

資料收集

5.1　Python 環境設置

5.1.1　為什麼需要設置環境？

在開始介紹爬蟲之前，需要先設置 Python 開發環境。

為什麼需要設置開發環境呢？一般開發上，不會只寫一套程式，可能開發多個專案、可能 clone 別人 GitHub 上的專案下來執行，因此，獨立的 Python 環境，有助於安裝 Python Package 時，不會影響其他專案、造成套件衝突等等。

5.1.2　Python 環境、套件管理工具 --- pipenv

本書中，將介紹 pipenv，pipenv 是 Python 官方推薦的套件管理工具：
https://packaging.python.org/guides/tool-recommendations/

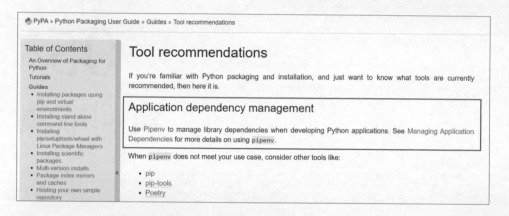

當然讀者也可以使用 conda、Virtualenv 等工具，pipenv 與前面幾個最大
的差別在於，會自動幫你建立 Python 虛擬環境，並自動管理 Python 使用
的 Package，自動 vs 手動，當然自動會比較好。

❏　安裝方式

首先，需要先安裝 pyenv。

```
sudo apt-get update
sudo apt install curl git bzip2 -y
curl https://pyenv.run | bash
```

環境設定

```
echo 'export LC_ALL=C.UTF-8' >> ~/.bashrc
echo 'export LANG=C.UTF-8' >> ~/.bashrc
echo 'export PYENV_ROOT="$HOME/.pyenv"' >> ~/.bashrc
echo 'export PATH="$PYENV_ROOT/shims:$PATH"' >> ~/.bashrc
echo 'export PATH="$PYENV_ROOT/bin:$PATH"' >> ~/.bashrc
echo -e 'if command -v pyenv 1>/dev/null 2>&1; then\n  eval "$(pyenv
init -)"\nfi' >> ~/.bashrc
exec $SHELL
```

使用 pyenv 安裝 miniconda，你可自行選擇版本安裝，這邊選擇
miniconda 3-4.3.30。

```
pyenv install miniconda3-4.3.30
```

設定 global Python 環境

```
pyenv global miniconda3-4.3.30
```

安裝 pipenv

```
pip install pipenv
```

當你使用 pipenv 時，就自動包含了獨立環境，以下將會介紹使用 pipenv
的方式。首先 pipenv 安裝套件方法與 pip 一樣，以 Flask 為例：

```
pipenv install flask==2.0.1
```

註：如果遇到 pipenv 無法找到對應的 python 路徑時，可以使用以下指
令，強制設定路徑。

```
pipenv --python ~/.pyenv/versions/miniconda3-4.3.30/bin/python
```

安裝成功畫面如下。

```
PROBLEMS    OUTPUT    DEBUG CONSOLE    TERMINAL

sam@DESKTOP-IKT69L5:~/FinMindBook$ pipenv install flask
Creating a Pipfile for this project…
Installing flask…
Adding flask to Pipfile's [packages]…
✓ Installation Succeeded
Pipfile.lock not found, creating…
Locking [dev-packages] dependencies…
Locking [packages] dependencies…
Building requirements...
Resolving dependencies...
✓ Success!
Updated Pipfile.lock (e239e5)!
Installing dependencies from Pipfile.lock (e239e5)…
🐍                                            0/0 — 00:00:00
To activate this project's virtualenv, run pipenv shell.
Alternatively, run a command inside the virtualenv with pipenv run.
sam@DESKTOP-IKT69L5:~/FinMindBook$ ▊
```

資料夾下，會出現兩個檔案：

Pipfile

```
✕     File  Edit  Selection  View  Go  Debug  Terminal  Help          Pipfile - book [WSL: Ubuntu-18.04] - Visual St

📄     EXPLORER                              ≡ Pipfile      ✕

       ∨ OPEN EDITORS                        ≡ Pipfile
🔍       ✕ ≡ Pipfile                          1   [[source]]
       ∨ BOOK [WSL: UBUNTU-18.04]            2   name = "pypi"
                                             3   url = "https://pypi.org/simple"
🔀       Pipfile                             4   verify_ssl = true
        {} Pipfile.lock         1            5
                                             6   [dev-packages]
🐞                                            7
                                             8   [packages]
                                             9   flask = "*"          2
🔲                                           10
                                            11   [requires]
🧩                                           12   python_version = "3.6"
                                            13
```

Pipfile.lock

```
≡ Pipfile          {} Pipfile.lock ×
{} Pipfile.lock > {} default > {} importlib-metadata > ▣ markers
  20              "hashes": [
  21                  "sha256:8c04c11192119b1ef78ea049e0a6f0463e4c48ef00a30160c704337586f3ad7a",
  22                  "sha256:fba402a4a47334742d782209a7c79bc448911afe1149d07bdabdf480b3e2f4b6"
  23              ],
  24              "markers": "python_version >= '3.6'",
  25              "version": "==8.0.1"
  26          },
  27          "dataclasses": {
  28              "hashes": [
  29                  "sha256:0201d89fa866f68c8ebd9d08ee6ff50c0b255f8ec63a71c16fda7af82bb887bf",
  30                  "sha256:8479067f342acf957dc82ec415d355ab5edb7e7646b90dc6e2fd1d96ad084c97"
  31              ],
  32              "markers": "python_version < '3.7'",
  33              "version": "==0.8"
  34          },
  35          "flask": {
  36              "hashes": [
  37                  "sha256:1c4c257b1892aec1398784c63791cbaa43062f1f7aeb555c4da961b20ee68f55",
  38                  "sha256:a6209ca15eb63fc9385f38e452704113d679511d9574d09b2cf9183ae7d20dc9"
  39              ],
  40              "index": "pypi",
  41              "version": "==2.0.1"
  42          },
```

在 Pipfile 中，packages 底下的 Flask，是剛剛安裝的 package，pipenv 會
幫你管理所使用的套件，requires 底下的 python_version 是目前使用的版
本。

Pipfile.lock 中，有 flask package 實際使用的版本號與 hashes。

接著，單純執行 Python 並 import flask，會出現：

ModuleNotFoundError: No module named 'flask'

```
sam@DESKTOP-IKT69L5:~/FinMindBook$ python
Python 3.6.3 |Anaconda, Inc.| (default, Oct 13 2017, 12:02:49)
[GCC 7.2.0] on linux
Type "help", "copyright", "credits" or "license" for more information.
im>>> import flask
Traceback (most recent call last):
  File "<stdin>", line 1, in <module>
ModuleNotFoundError: No module named 'flask'
>>>
```

這時你可能會有疑問，上面不是安裝了嗎？為什麼沒有 package，這就是 pipenv 的其中一個特點，自動建立獨立環境，剛剛安裝 flask 時，是安裝到 pipenv 建立的獨立環境中，不會安裝到系統環境，所以這時 import 才會失敗。那要如何啟動 pipenv 獨立環境呢？以下指令即可開啟獨立環境：

```
pipenv run python
```

pipenv 進入獨立環境、並啟動 Python，兩個動作包在一起，接著再執行，import flask，這時就正常了。

```
sam@DESKTOP-IKT69L5:~/FinMindBook$ pipenv run python
Creating a Pipfile for this project…
Python 3.6.9 (default, Jan 26 2021, 15:33:00)
[GCC 8.4.0] on linux
Type "help", "copyright", "credits" or "license" for more information.
>>> import flask
>>>
```

由於 pipenv 會比較套件之間的相依性，網路上有不少使用者反應，pipenv 很慢，因此，筆者再介紹另一個知名的套件管理工具 - poetry。

5.1.3 Python 環境、套件管理工具 --- poetry

當一個專案發展到越來越大時，一定會遇到，套件之間相依性問題，pipenv 就會越跑越久，這時，讀者可以使用本節工具 - poetry。

在筆者撰寫第二版當下（2023-03），Python 官方網站做了更動，將 poetry 與 pipenv 放到同一段文字敘述中，讀者可以比較 5.1.2 的圖。https://packaging.python.org/guides/tool-recommendations/

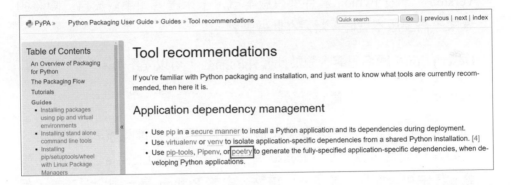

這也說明了，poetry 這項工具的可信度。新的工具很多，身為初學者，當然優先學習官方推薦的工具。

筆者將一一介紹，如何使用 poetry。

安裝 poetry

```
pip install poetry==1.1.15
```

初始化

```
poetry init
```

初始化過程中，poetry 會詢問許多問題，這時讀者只需按 enter 一鍵到底即可，如下圖筆者在此也一一介紹各個問題的含意：

- Package name：poetry 在初始化一個 package，因此需要名稱，如 Python 知名專案 pandas、numpy、flask 等，在此只需填入讀者專案的名稱，以便後續 import 使用。筆者在此只是做個範例，因此直接 enter 下一步即可。

- Version：一般 Python 套件都有版本號，未來讀者如果對於自己開發的專案，需要版本號時，就是在此做調整。

- Description：顧名思義，對於此套件的描述，可以先不填。

- Author：該套件的作者。

- License：許可證，如果需要發布這個專案，可以選擇開源專案的程度，例如是否商用，常見的 License 有 MIT、Apache License 2.0、BSD 等等。其中可否商用，算是一大重點，由於 Python 是一個開源社群，但開源不一定商用，可能只提供非商業用途，讀者在使用時，可以觀察是否有 MIT、Apache License 2.0、BSD 這幾個常見的商用許可。

- Compatible Python versions：Python 的版本，讀者可根據需求選用 3.6 ～ 3.9。後續選項相對不重要，只需下一步即可。

```
sam@DESKTOP-IKT69L5:~/FinMindBook/DataEngineering/Chapter5/5.1/5.1.3$ poetry init

This command will guide you through creating your pyproject.toml config.

Package name [5.1.3]:
Version [0.1.0]:
Description []:
Author [sam <samlin266118@gmail.com>, n to skip]:
License []:
Compatible Python versions [^3.6]:

Would you like to define your main dependencies interactively? (yes/no) [yes]
You can specify a package in the following forms:
  - A single name (requests)
  - A name and a constraint (requests@^2.23.0)
  - A git url (git+https://github.com/python-poetry/poetry.git)
  - A git url with a revision (git+https://github.com/python-poetry/poetry.git#develop)
  - A file path (../my-package/my-package.whl)
  - A directory (../my-package/)
  - A url (https://example.com/packages/my-package-0.1.0.tar.gz)
Search for package to add (or leave blank to continue):

Would you like to define your development dependencies interactively? (yes/no) [yes]
Search for package to add (or leave blank to continue):

Generated file

[tool.poetry]
name = "5.1.3"
version = "0.1.0"
description = ""
authors = ["sam <samlin266118@gmail.com>"]

[tool.poetry.dependencies]
python = "^3.6"

[tool.poetry.dev-dependencies]

[build-system]
requires = ["poetry-core>=1.0.0"]
build-backend = "poetry.core.masonry.api"


Do you confirm generation? (yes/no) [yes]
```

完成初始化後,與 pipenv 類似,會生成一個設定檔,pyproject.toml,接著筆者來安裝套件。

當使用 poetry 時,與 pipenv 類似,也包含了獨立環境建立,同樣以安裝 flask 為例。

```
poetry add flask==2.0.1
```

安裝成功畫面如下：

```
sam@DESKTOP-IKT69L5:~/FinMindBook/DataEngineering/Chapter5/5.1/5.1.3$ poetry add flask==2.0.1

Updating dependencies
Resolving dependencies... (44.4s)

Writing lock file

Package operations: 10 installs, 0 updates, 0 removals

  • Installing typing-extensions (4.1.1)
  • Installing zipp (3.6.0)
  • Installing dataclasses (0.8)
  • Installing importlib-metadata (4.8.3)
  • Installing markupsafe (2.0.1)
  • Installing click (8.0.4)
  • Installing itsdangerous (2.0.1)
  • Installing jinja2 (3.0.3)
  • Installing werkzeug (2.0.3)
  • Installing flask (2.0.1)
sam@DESKTOP-IKT69L5:~/FinMindBook/DataEngineering/Chapter5/5.1/5.1.3$ █
```

那要如何與 pipenv 一樣，開啟獨立環境呢？

執行

```
poetry run python
```

```
sam@DESKTOP-IKT69L5:~/FinMindBook/DataEngineering/Chapter5/5.1/5.1.3$ poetry run python
Python 3.6.3 |Anaconda, Inc.| (default, Oct 13 2017, 12:02:49)
[GCC 7.2.0] on linux
Type "help", "copyright", "credits" or "license" for more information.
>>> import flask
>>> flask.__version__
'2.0.1'
>>> █
```

以上，就完成了 Python 環境設置，基本上 pipenv or poetry 使用上都大同小異，讀者可自行選用適合自己的工具，本書中主要以 pipenv 為例。

下一個章節，將介紹爬蟲、資料庫安裝與分散式。

5.2　爬蟲

5.2.1　什麼是爬蟲？

爬蟲是使用程式，自動化的方式，擷取特定網站資訊，收集資料的方式。而一般網站，為了防範程式暴力爬蟲，可能會設置驗證碼，或是一定時間內大量連續訪問，會禁止你連線一段時間，以下本書將以證交所、櫃買中心和期交所爬蟲為範例。

程式架構如下，Pipfile 就是上面介紹的套件管理工具，src 底下的 taifex_crawler.py、tpex_crawler.py、twse_crawler.py 分別是期交所、櫃買中心、證交所爬蟲。

https://github.com/FinMind/FinMindBook/tree/master/DataEngineering/Chapter5/5.2

```
├── Pipfile
├── Pipfile.lock
├── README.md
└── src
    ├── taifex_crawler.py
    ├── tpex_crawler.py
    └── twse_crawler.py
```

5.2.2　證交所爬蟲範例

https://www.twse.com.tw/zh/page/trading/exchange/MI_INDEX.html

證交所網址如上，以股價為例，在做爬蟲之前，需要觀察在網頁上操作的過程中，對網頁做了什麼樣的動作，並用程式模仿這段動作，就可以取得想要的資訊。

1. 在網站上點擊右鍵,選取**檢查**,檢查可以幫助觀察,當查詢股價時,實際上發什麼 request 請求給證交所伺服器,之後再用程式去模擬這段 request。

2. 選取 **Network**,Network 包含了接下來對伺服器發送的 request 與 response,讀者需要從這觀察,請求 request 與回應 response。

3. 選擇全部,這是要爬取的股價分類。

4. 點選查詢之後，右邊 Network 部分，會出現伺服器的回應，選取第一個 **MI_INDEX?response=json**，觀察查詢動作，是針對哪個網址發送 request，發送內容是什麼，收到的 response 是不是股價資訊要的。

5. 點選 **Headers**，觀察到查詢這個動作，是對

   https://www.twse.com.tw/exchangeReport/MI_INDEX?response=json&date=20210309&type=ALL

   發送 request 請求。

   這裡的 **GET**，是 request 的方式，有些網站是 **POST**，這要特別注意。

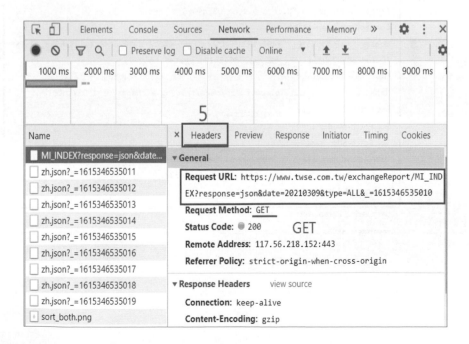

6. 觀察 header，待會書中下面的 Python Code 會用到，在模擬瀏覽器登入時所帶的參數。

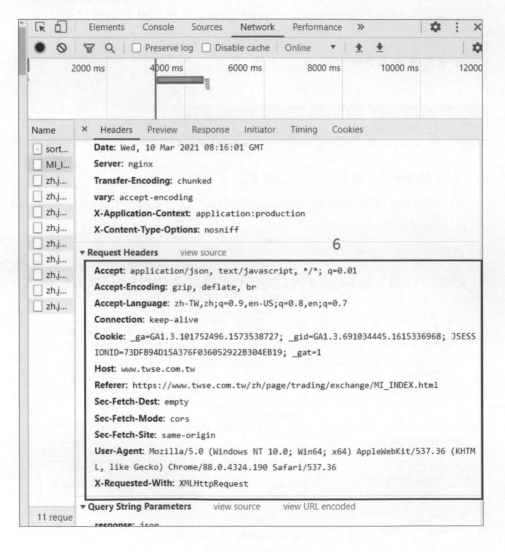

7. 點選 **Preview**，這裡可以看到對網站發送請求時得到的 data，這裡有 data1 到 data9，那怎麼判斷，哪些是股價資訊呢？

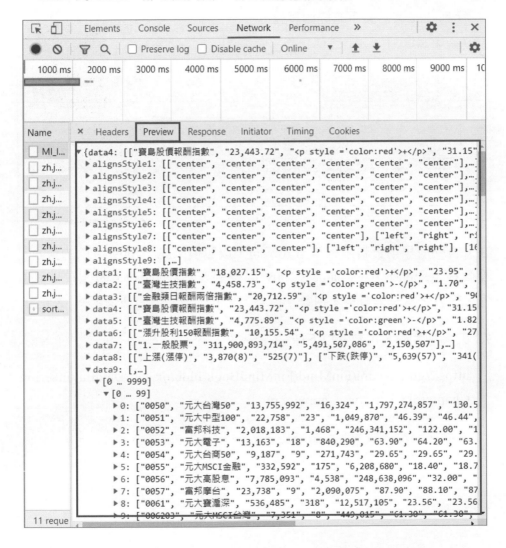

8. 把網頁滾輪滑到最下面，到每日**收盤行情**的位置，接著來比對一下 **Preview**，觀察出，**data9** 是所需要的股價資料，欄位名稱在 **fields9**，找到 data 後，接下來開始爬蟲。

				(元股)									(元交易單位)				
證券代號	證券名稱	成交股數	成交筆數	成交金額	開盤價	最高價	最低價	收盤價	漲跌(+/-)	漲跌價差		最後揭示買價	最後揭示買量	最後揭示賣價	最後揭示賣量	本益比
0050	元大台灣50	13,755,992	16,324	1,797,274,857	130.55	131.60	129.75	131.35	-	0.50		131.35	9	131.40	16	0.00
0051	元大中型100	22,758	23	1,049,870	46.39	46.44	45.74	46.20	-	0.19		46.35	3	46.36	1	0.00
0052	富邦科技	2,018,183	1,468	246,341,152	122.00	123.15	121.30	123.00	-	0.85		122.95	1	123.00	9	0.00
0053	元大電子	13,163	18	840,290	63.90	64.20	63.60	64.00	-	0.30		63.90	1	64.30	10	0.00
0054	元大台商50	9,187	9	271,743	29.65	29.65	29.40	29.60	-	0.45		29.54	55	29.70	1	0.00
0055	元大MSCI金融	332,592	175	6,208,680	18.40	18.77	18.40	18.75	+ 7	0.44		18.72	1	18.75	8	0.00
0056	元大高股息	7,785,093	4,538	248,638,096	32.00	32.01	31.84	32.00	-	0.00		31.99	1	32.00	259	0.00
0057	富邦摩台	23,738	9	2,090,075	87.90	88.10	87.85	88.10	-	0.55		88.05	10	88.50	10	0.00
0061	元大寶滬深	536,485	318	12,517,105	23.56	23.56	22.95	23.42	-	0.29		23.41	6	23.42	49	0.00
006203	元大MSCI台灣	7,351	8	449,015	61.30	61.30	60.80	61.30	-	0.30		61.10	3	61.35	35	0.00

110年03月09日 每日收盤行情(全部)　每頁 10 ∨ 筆　上頁 1 2 3 4 5 … 2149 下頁

註：證交所早期資料跟近年格式不同，以股價來說，2009 以前的資料，股價在 response 中的 data8，2009 以後的資料在 data9，因此在做爬蟲時，需要多看幾筆資料來判斷，以下的程式碼會處理這種情境。

9. 爬蟲程式碼，可參考以下連結：

https://github.com/FinMind/FinMindBook/blob/master/DataEngineering/Chapter5/5.2/src/twse_crawler.py

在開始之前，你需要先執行以下指令，安裝 Package。

```
pipenv install pandas requests pydantic loguru
```

爬蟲程式碼：

```
src/twse_crawler.py
import datetime
import sys
import time
import typing

import pandas as pd
import requests
from loguru import logger
from pydantic import BaseModel


def clear_data(
    df: pd.DataFrame,
) -> pd.DataFrame:
    """資料清理，將文字轉成數字"""
    df["Dir"] = (
        df["Dir"]
        .str.split(">")
        .str[1]
        .str.split("<")
        .str[0]
    )
    df["Change"] = (
        df["Dir"] + df["Change"]
    )
    df["Change"] = (
        df["Change"]
        .str.replace(" ", "")
        .str.replace("X", "")
```

```python
        .astype(float)
    )
    df = df.fillna("")
    df = df.drop(["Dir"], axis=1)
    for col in [
        "TradeVolume",
        "Transaction",
        "TradeValue",
        "Open",
        "Max",
        "Min",
        "Close",
        "Change",
    ]:
        df[col] = (
            df[col]
            .astype(str)
            .str.replace(",", "")
            .str.replace("X", "")
            .str.replace("+", "")
            .str.replace("----", "0")
            .str.replace("---", "0")
            .str.replace("--", "0")
        )
    return df


def colname_zh2en(
    df: pd.DataFrame,
    colname: typing.List[str],
) -> pd.DataFrame:
    """資料欄位轉換，英文有助於接下來存入資料庫"""
```

```python
    taiwan_stock_price = {
        "證券代號": "StockID",
        "證券名稱": "",
        "成交股數": "TradeVolume",
        "成交筆數": "Transaction",
        "成交金額": "TradeValue",
        "開盤價": "Open",
        "最高價": "Max",
        "最低價": "Min",
        "收盤價": "Close",
        "漲跌(+/-)": "Dir",
        "漲跌價差": "Change",
        "最後揭示買價": "",
        "最後揭示買量": "",
        "最後揭示賣價": "",
        "最後揭示賣量": "",
        "本益比": "",
    }
    df.columns = [
        taiwan_stock_price[col]
        for col in colname
    ]
    df = df.drop([""], axis=1)
    return df


def twse_header():
    """網頁瀏覽時, 所帶的 request header 參數, 模仿瀏覽器發送 request"""
    return {
        "Accept": "application/json, text/javascript, */*; q=0.01",
        "Accept-Encoding": "gzip, deflate",
        "Accept-Language": "zh-TW,zh;q=0.9,en-US;q=0.8,en;q=0.7",
```

```python
        "Connection": "keep-alive",
        "Host": "www.twse.com.tw",
        "Referer": "https://www.twse.com.tw/zh/page/trading/exchange/
MI_INDEX.html",
        "User-Agent": "Mozilla/5.0 (Windows NT 10.0; Win64;
x64) AppleWebKit/537.36 (KHTML, like Gecko) Chrome/71.0.3578.98
Safari/537.36",
        "X-Requested-With": "XMLHttpRequest",
    }


def crawler_twse(
    date: str,
) -> pd.DataFrame:
    """
    證交所網址
    https://www.twse.com.tw/zh/page/trading/exchange/MI_INDEX.html
    """
    # headers 中的 Request url
    url = (
        "https://www.twse.com.tw/exchangeReport/MI_INDEX"
        "?response=json&date={date}&type=ALL"
    )
    url = url.format(
        date=date.replace("-", "")
    )
    # 避免被證交所 ban ip, 在每次爬蟲時, 先 sleep 5 秒
    time.sleep(5)
    # request method
    res = requests.get(
        url, headers=twse_header()
    )
```

```
if (
    res.json()["stat"]
    == "很抱歉，沒有符合條件的資料!"
):
    # 如果 date 是周末，會回傳很抱歉，沒有符合條件的資料!
    return pd.DataFrame()
# 2009 年以後的資料，股價在 response 中的 data9
# 2009 年以後的資料，股價在 response 中的 data8
# 不同格式，在證交所的資料中，是很常見的，
# 沒資料的情境也要考慮進去，例如現在週六沒有交易，但在 2007 年週六是有
交易的
try:
    if "data9" in res.json():
        df = pd.DataFrame(
            res.json()["data9"]
        )
        colname = res.json()[
            "fields9"
        ]
    elif "data8" in res.json():
        df = pd.DataFrame(
            res.json()["data8"]
        )
        colname = res.json()[
            "fields8"
        ]
    elif res.json()["stat"] in [
        "查詢日期小於93年2月11日，請重新查詢!",
        "很抱歉，沒有符合條件的資料!",
    ]:
        return pd.DataFrame()
except BaseException:
```

```
        return pd.DataFrame()

    if len(df) == 0:
        return pd.DataFrame()
    # 欄位中英轉換
    df = colname_zh2en(
        df.copy(), colname
    )
    df["date"] = date
    return df


class TaiwanStockPrice(BaseModel):
    StockID: str
    TradeVolume: int
    Transaction: int
    TradeValue: int
    Open: float
    Max: float
    Min: float
    Close: float
    Change: float
    date: str


def check_schema(
    df: pd.DataFrame,
) -> pd.DataFrame:
    """檢查資料型態，確保每次要上傳資料庫前，型態正確"""
    df_dict = df.to_dict("records")
    df_schema = [
        TaiwanStockPrice(**dd).__dict__
```

```python
        for dd in df_dict
    ]
    df = pd.DataFrame(df_schema)
    return df


def gen_date_list(
    start_date: str, end_date: str
) -> typing.List[str]:
    """建立時間列表, 用於爬取所有資料"""
    start_date = (
        datetime.datetime.strptime(
            start_date, "%Y-%m-%d"
        ).date()
    )
    end_date = (
        datetime.datetime.strptime(
            end_date, "%Y-%m-%d"
        ).date()
    )
    days = (
        end_date - start_date
    ).days + 1
    date_list = [
        str(
            start_date
            + datetime.timedelta(
                days=day
            )
        )
        for day in range(days)
    ]
```

```python
    return date_list


def main(
    start_date: str, end_date: str
):
    """證交所寫明，※ 本資訊自民國93年2月11日起提供"""
    date_list = gen_date_list(
        start_date, end_date
    )
    for date in date_list:
        logger.info(date)
        df = crawler_twse(date)
        if len(df) > 0:
            # 資料清理
            df = clear_data(df.copy())
            # 檢查資料型態
            df = check_schema(df.copy())
            # 這邊先暫時存成 file，下個章節將會上傳資料庫
            df.to_csv(
                f"taiwan_stock_price_twse_{date}.csv",
                index=False,
            )


if __name__ == "__main__":
    start_date, end_date = sys.argv[1:]
    main(start_date, end_date)
```

執行以下指令，即可爬取 2021-03-20 到 2021-03-25 的證交所股價。

```
pipenv run python src/twse_crawler.py 2021-03-20 2021-03-25
```

基本上證交所其他資料的格式,如三大法人、融資融券,都與股價相同,因此模仿上面的程式碼,都可以正常的爬取資料,但證交所只有上市資料,而平常股市分析,會使用上市與上櫃資料,接下來,介紹櫃買中心爬蟲。

5.2.3 櫃買中心爬蟲範例

https://www.tpex.org.tw/web/stock/aftertrading/otc_quotes_no1430/stk_wn1430.php?l=zh-tw

櫃買中心網址如上,爬蟲方式,基本上跟證交所一樣,就不一步步贅述:

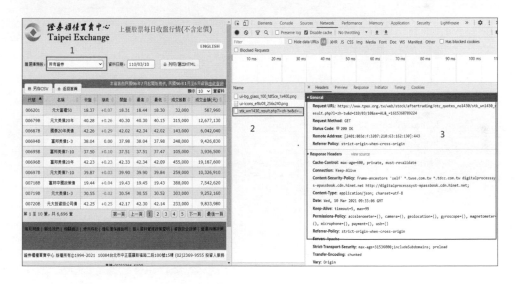

爬蟲程式碼,可參考以下連結:

https://github.com/FinMind/FinMindBook/blob/master/DataEngineering/Chapter5/5.2/src/tpex_crawler.py

```
src/tpex_crawler.py
import datetime
import sys
import time
import typing

import pandas as pd
import requests
from loguru import logger
from pydantic import BaseModel


def clear_data(
    df: pd.DataFrame,
) -> pd.DataFrame:
    """資料清理, 將文字轉成數字"""
    for col in [
        "TradeVolume",
        "Transaction",
        "TradeValue",
        "Open",
        "Max",
        "Min",
        "Close",
        "Change",
    ]:
        df[col] = (
            df[col]
            .astype(str)
            .str.replace(",", "")
            .str.replace("X", "")
            .str.replace("+", "")
            .str.replace("----", "0")
            .str.replace("---", "0")
```

```python
            .str.replace("--", "0")
            .str.replace(" ", "")
            .str.replace("除權息", "0")
            .str.replace("除息", "0")
            .str.replace("除權", "0")
        )
    return df


def set_column(
    df: pd.DataFrame,
) -> pd.DataFrame:
    """設定資料欄位名稱"""
    df.columns = [
        "StockID",
        "Close",
        "Change",
        "Open",
        "Max",
        "Min",
        "TradeVolume",
        "TradeValue",
        "Transaction",
    ]
    return df


def tpex_header():
    """網頁瀏覽時, 所帶的 request header 參數, 模仿瀏覽器發送 request"""
    return {
        "Accept": "application/json, text/javascript, */*; q=0.01",
        "Accept-Encoding": "gzip, deflate",
        "Accept-Language": "zh-TW,zh;q=0.9,en-US;q=0.8,en;q=0.7",
        "Connection": "keep-alive",
```

```
        "Host": "www.tpex.org.tw",
        "Referer": "https://www.tpex.org.tw/web/stock/aftertrading/
otc_quotes_no1430/stk_wn1430.php?l=zh-tw",
        "User-Agent": "Mozilla/5.0 (Windows NT 10.0; Win64; x64)
AppleWebKit/537.36 (KHTML, like Gecko) Chrome/73.0.3683.103
Safari/537.36",
        "X-Requested-With": "XMLHttpRequest",
    }


def convert_date(date: str) -> str:
    year, month, day = date.split("-")
    year = int(year) - 1911
    return f"{year}/{month}/{day}"


def crawler_tpex(
    date: str,
) -> pd.DataFrame:
    """
    櫃買中心網址
    https://www.tpex.org.tw/web/stock/aftertrading/otc_quotes_no1430/
stk_wn1430.php?l=zh-tw
    """
    # headers 中的 Request url
    url = (
        "https://www.tpex.org.tw/web/stock/aftertrading/"
        "otc_quotes_no1430/stk_wn1430_result.php?"
        "l=zh-tw&d={date}&se=AL"
    )
    url = url.format(
        date=convert_date(date)
    )
    # 避免被櫃買中心 ban ip，在每次爬蟲時，先 sleep 5 秒
```

```
    time.sleep(5)
    # request method
    res = requests.get(
        url, headers=tpex_header()
    )
    data = res.json().get("aaData", "")
    if not data:
        return pd.DataFrame()
    df = pd.DataFrame(data)

    if len(df) == 0:
        return pd.DataFrame()
    # 櫃買中心回傳的資料，並無資料欄位，因此這裡直接用 index 取特定欄位
    df = df[[0, 2, 3, 4, 5, 6, 7, 8, 9]]
    # 欄位中英轉換
    df = set_column(df.copy())
    df["date"] = date
    return df


class TaiwanStockPrice(BaseModel):
    StockID: str
    TradeVolume: int
    Transaction: int
    TradeValue: int
    Open: float
    Max: float
    Min: float
    Close: float
    Change: float
    date: str


def check_schema(
```

```python
    df: pd.DataFrame,
) -> pd.DataFrame:
    """檢查資料型態，確保每次要上傳資料庫前，型態正確"""
    df_dict = df.to_dict("records")
    df_schema = [
        TaiwanStockPrice(**dd).__dict__
        for dd in df_dict
    ]
    df = pd.DataFrame(df_schema)
    return df


def gen_date_list(
    start_date: str, end_date: str
) -> typing.List[str]:
    """建立時間列表，用於爬取所有資料"""
    start_date = (
        datetime.datetime.strptime(
            start_date, "%Y-%m-%d"
        ).date()
    )
    end_date = (
        datetime.datetime.strptime(
            end_date, "%Y-%m-%d"
        ).date()
    )
    days = (
        end_date - start_date
    ).days + 1
    date_list = [
        str(
            start_date
            + datetime.timedelta(
                days=day
```

```python
            )
        )
        for day in range(days)
    ]
    return date_list


def main(
    start_date: str, end_date: str
):
    """櫃買中心寫明，本資訊自民國96年7月起開始提供"""
    date_list = gen_date_list(
        start_date, end_date
    )
    for date in date_list:
        logger.info(date)
        df = crawler_tpex(date)
        if len(df) > 0:
            # 資料清理
            df = clear_data(df.copy())
            # 檢查資料型態
            df = check_schema(df.copy())
            # 這邊先暫時存成 file，下個章節將會上傳資料庫
            df.to_csv(
                f"taiwan_stock_price_tpex_{date}.csv",
                index=False,
            )


if __name__ == "__main__":
    start_date, end_date = sys.argv[1:]
    main(start_date, end_date)
```

執行以下指令，即可爬取 2021-03-20 到 2021-03-25 的櫃買中心股價。

```
pipenv run python src/tpex_crawler.py 2021-03-20 2021-03-25
```

與證交所相同，櫃買中心的其他資料格式也一致，本書到此，對於證交所、櫃買中心的爬蟲，可以說是掌握八成了，以下將介紹期交所爬蟲。

5.2.4 期交所爬蟲範例

https://www.taifex.com.tw/cht/3/futDailyMarketView

期交所網址如上，跟證交所、櫃買中心一樣的方法，另外期貨與選擇權相同，這邊以期貨為例：

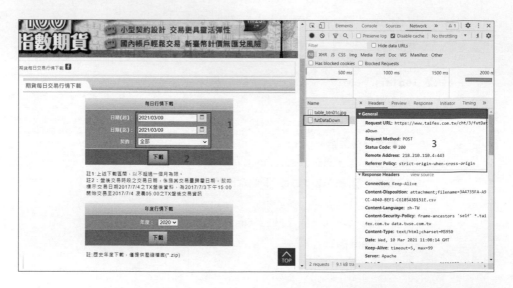

這邊稍微不同，是用 POST，需要將右邊的滑輪，拉到最下面，觀察 POST 發送什麼 Form Data。

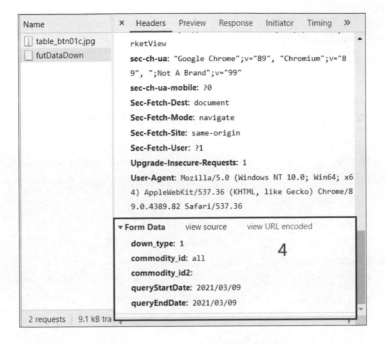

接著以下的程式碼，會模仿瀏覽器，發送 POST 給期交所伺服器。

可參考以下連結：

https://github.com/FinMind/FinMindBook/blob/master/DataEngineering/
Chapter5/5.2/src/taifex_crawler.py

```
src/taifex_crawler.py
import datetime
import io
import sys
import time
import typing


import pandas as pd
import requests
```

```python
from loguru import logger
from pydantic import BaseModel


def futures_header():
    """網頁瀏覽時, 所帶的 request header 參數, 模仿瀏覽器發送 request"""
    return {
        "Accept": "text/html,application/xhtml+xml,application/xml;
q=0.9,image/webp,image/apng,*/*;q=0.8,application/signed-exchange;
v=b3;q=0.9",
        "Accept-Encoding": "gzip, deflate, br",
        "Accept-Language": "zh-TW,zh;q=0.9,en-US;q=0.8,en;q=0.7",
        "Cache-Control": "no-cache",
        "Connection": "keep-alive",
        "Content-Length": "101",
        "Content-Type": "application/x-www-form-urlencoded",
        "Host": "www.taifex.com.tw",
        "Origin": "https://www.taifex.com.tw",
        "Pragma": "no-cache",
        "Referer": "https://www.taifex.com.tw/cht/3/dlFutDailyMarketView",
        "Sec-Fetch-Dest": "document",
        "Sec-Fetch-Mode": "navigate",
        "Sec-Fetch-Site": "same-origin",
        "Sec-Fetch-User": "?1",
        "Upgrade-Insecure-Requests": "1",
        "User-Agent": "Mozilla/5.0 (Windows NT 10.0; Win64; x64)
AppleWebKit/537.36 (KHTML, like Gecko) Chrome/81.0.4044.113 Safari/537.36",
    }


def colname_zh2en(
    df: pd.DataFrame,
```

```python
) -> pd.DataFrame:
    """資料欄位轉換，英文有助於接下來存入資料庫"""
    colname_dict = {
        "交易日期": "date",
        "契約": "FuturesID",
        "到期月份(週別)": "ContractDate",
        "開盤價": "Open",
        "最高價": "Max",
        "最低價": "Min",
        "收盤價": "Close",
        "漲跌價": "Change",
        "漲跌%": "ChangePer",
        "成交量": "Volume",
        "結算價": "SettlementPrice",
        "未沖銷契約數": "OpenInterest",
        "交易時段": "TradingSession",
    }
    df = df.drop(
        [
            "最後最佳買價",
            "最後最佳賣價",
            "歷史最高價",
            "歷史最低價",
            "是否因訊息面暫停交易",
            "價差對單式委託成交量",
        ],
        axis=1,
    )
    df.columns = [
        colname_dict[col]
        for col in df.columns
    ]
```

```python
    return df


def clean_data(
    df: pd.DataFrame,
) -> pd.DataFrame:
    """資料清理"""
    df["date"] = df["date"].str.replace(
        "/", "-"
    )
    df["ChangePer"] = df[
        "ChangePer"
    ].str.replace("%", "")
    df["ContractDate"] = (
        df["ContractDate"]
        .astype(str)
        .str.replace(" ", "")
    )
    if "TradingSession" in df.columns:
        df["TradingSession"] = df[
            "TradingSession"
        ].map(
            {
                "一般": "Position",
                "盤後": "AfterMarket",
            }
        )
    else:
        df[
            "TradingSession"
        ] = "Position"
    for col in [
```

```
        "Open",
        "Max",
        "Min",
        "Close",
        "Change",
        "ChangePer",
        "Volume",
        "SettlementPrice",
        "OpenInterest",
    ]:
        df[col] = (
            df[col]
            .replace("-", "0")
            .astype(float)
        )
    df = df.fillna(0)
    return df


def crawler_futures(
    date: str,
) -> pd.DataFrame:
    """期交所爬蟲"""
    url = "https://www.taifex.com.tw/cht/3/futDataDown"
    form_data = {
        "down_type": "1",
        "commodity_id": "all",
        "queryStartDate": date.replace(
            "-", "/"
        ),
        "queryEndDate": date.replace(
            "-", "/"
```

```
        ),
    }
    # 避免被期交所 ban ip, 在每次爬蟲時, 先 sleep 5 秒
    time.sleep(5)
    resp = requests.post(
        url,
        headers=futures_header(),
        data=form_data,
    )
    if resp.ok:
        if resp.content:
            df = pd.read_csv(
                io.StringIO(
                    resp.content.decode(
                        "big5"
                    )
                ),
                index_col=False,
            )
    else:
        return pd.DataFrame()
    return df


class TaiwanFuturesDaily(BaseModel):
    date: str
    FuturesID: str
    ContractDate: str
    Open: float
    Max: float
    Min: float
    Close: float
```

```
    Change: float

    ChangePer: float

    Volume: float

    SettlementPrice: float

    OpenInterest: int

    TradingSession: str


def check_schema(

    df: pd.DataFrame,

) -> pd.DataFrame:

    """檢查資料型態, 確保每次要上傳資料庫前, 型態正確"""

    df_dict = df.to_dict("records")

    df_schema = [

        TaiwanFuturesDaily(

            **dd

        ).__dict__

        for dd in df_dict

    ]

    df = pd.DataFrame(df_schema)

    return df


def gen_date_list(

    start_date: str, end_date: str

) -> typing.List[str]:

    """建立時間列表, 用於爬取所有資料"""

    start_date = (

        datetime.datetime.strptime(

            start_date, "%Y-%m-%d"

        ).date()

    )
```

```python
    end_date = (
        datetime.datetime.strptime(
            end_date, "%Y-%m-%d"
        ).date()
    )
    days = (
        end_date - start_date
    ).days + 1
    date_list = [
        str(
            start_date
            + datetime.timedelta(
                days=day
            )
        )
        for day in range(days)
    ]
    return date_list


def main(
    start_date: str, end_date: str
):
    date_list = gen_date_list(
        start_date, end_date
    )
    for date in date_list:
        logger.info(date)
        df = crawler_futures(date)
        if len(df) > 0:
            # 欄位中英轉換
            df = colname_zh2en(
```

```
        df.copy()
    )
    # 資料清理
    df = clean_data(df.copy())
    # 檢查資料型態
    df = check_schema(df.copy())
    # 這邊先暫時存成 file，下個章節將會上傳資料庫
    df.to_csv(
        f"taiwan_futures_price_{date}.csv",
        index=False,
    )


if __name__ == "__main__":
    start_date, end_date = sys.argv[1:]
    main(start_date, end_date)
```

執行以下指令，即可爬取 2021-03-20 到 2021-03-25 的期交所期貨價格。

```
pipenv run python src/taifex_crawler.py 2021-03-20 2021-03-25
```

以上，台股資料，大多都能成功爬蟲了，當然爬蟲難度不高，問題是如何管理大量的爬蟲？如果將證交所、櫃買中心、期交所大多的資料都爬下來，可能有 50 隻爬蟲程式，那要如何管理？如果都存成 csv，那這些檔案要如何管理？本書將會在 5.3、5.4 章節，介紹爬蟲後上傳資料庫，在 5.5 章節，介紹 Celery 分散式爬蟲。在這之前，先架設資料庫，並將以上爬蟲資料存入資料庫。

5.3 資料庫架設

5.3.1 什麼是資料庫？

為什麼需要資料庫？可能剛開始接觸爬蟲的工程師，不太清楚資料庫的用途，這時本書用爬蟲，作為資料庫應用的場景。

資料庫是協助儲存、管理各種資料的工具，以股價為例，一年約股市交易約 200 天，當你爬蟲收集十年的歷史股價時，不太可能直接存成 Excel，即使你真的把全部股價存成一個 Excel，那除了股價，還有三大法人、融資券、期貨、選擇權等等的資料，直接存成 Excel 拓展性低，用資料庫來管理，會比較有效率，以下將會介紹如何用 Docker 一鍵架設資料庫。

5.3.2 關聯式資料庫 --- MySQL

在介紹之前，先來安裝 MySQL，將以下程式碼，存成 mysql.yml，或是可以從以下連結下載：

https://github.com/FinMind/FinMindBook/blob/master/DataEngineering/Chapter5/5.3/mysql.yml

```
mysql.yml
version: '3.3'
services:

  mysql:
    image: mysql:8.0
    # 設定 mysql 使用原生認證的密碼 hash
    command: mysqld --default-authentication-plugin=mysql_native_password
    ports:
```

```
        # docker publish port 3306 to 3306
        # (將 docker 內部 ip 3306, 跟外部 3306 做連結)
          - 3306:3306
        # - target: 3306
        #   published: 3306
        #   mode: host
      environment: # 環境變數, 設置 db, user, password, root_password
          MYSQL_DATABASE: mydb
          MYSQL_USER: user
          MYSQL_PASSWORD: test
          MYSQL_ROOT_PASSWORD: test
      volumes:
          - mysql:/var/lib/mysql
        # share volumes of docker container to outside volume,
        # let data persist
        # 將 docker container 的 volumes 跟外部 volumes 做連結,
        # 讓資料不會因為, container close or update, 而遺失
      networks:
          - dev


  phpmyadmin:
      image: phpmyadmin/phpmyadmin:5.1.0
      links:
          - mysql:db
      ports:
          - 8000:80
      depends_on:
        - mysql
      networks:
          - dev

networks:
```

```
    dev:

volumes:
  mysql:
    external: true
```

先建立 volume，使用以下指令：

```
docker volume create mysql
```

volume 只要建立一次就好，這是確保資料持久性，不會因為 Docker 關閉
而資料遺失，接著使用以下指令，就架好 MySQL 了。

```
docker-compose -f mysql.yml up
```

這時 Docker 會自動去網路上，下載 Image，如下圖：

```
sam@DESKTOP-IKT69L5:~/FinMindBook/DataEngineering/Chapter5/5.3$ docker-compose -f mysql.yml up
Pulling mysql (mysql:8.0)...
8.0: Pulling from library/mysql
b380bbd43752: Pull complete
f23cbf2ecc5d: Pull complete
30cfc6c29c0a: Pull complete
b38609286cbe: Pull complete
8211d9e66cd6: Pull complete
2313f9eeca4a: Pull complete
7eb487d00da0: Pull complete
4d7421c8152e: Pull complete
77f3d8811a28: Pull complete
cce755338cba: Pull complete
69b753046b9f: Pull complete
b2e64b0ab53c: Pull complete
Digest: sha256:6d7d4524463fe6e2b893ffc2b89543c81dec7ef82fb2020a1b27606666464d87
Status: Downloaded newer image for mysql:8.0
Pulling phpmyadmin (phpmyadmin/phpmyadmin:5.1.0)...
5.1.0: Pulling from phpmyadmin/phpmyadmin
45b42c59be33: Downloading [====================================>        ]  20.38MB/27.1MB
a48991d6909c: Download complete
935e2abd2c2c: Downloading [=======>                                     ]  12.88MB/76.68MB
61ccf45ccdb9: Waiting
27b5ac70765b: Waiting
5638b69045ba: Waiting
0fdaed064166: Waiting
e932cec09ced: Waiting
```

下載完畢之後,打開瀏覽器,輸入 http://localhost:8000/,帳號密碼是
root/test:

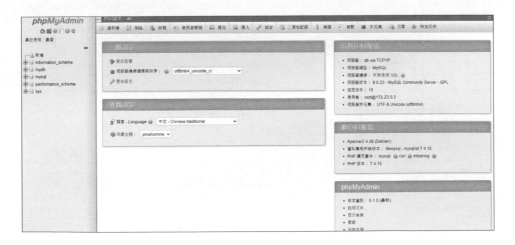

登入後,會看到以下畫面:

這時有 UI 可以操作，就可以開始使用資料庫了，先建立 financialdata 資料庫，方便接下來爬蟲使用，步驟如下。

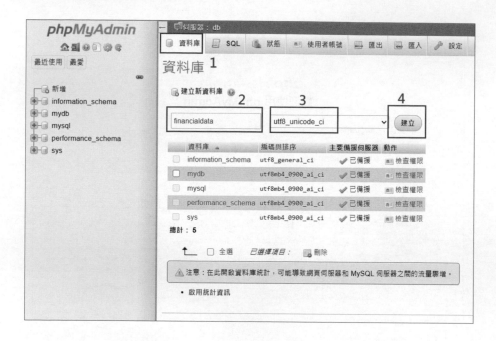

成功建立後，畫面如下：

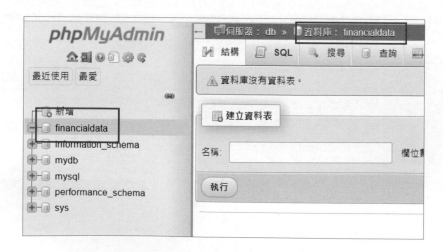

再回來看你剛剛的 terminal，會呈現以下樣子，卡住，無法做其他事：

```
PROBLEMS  OUTPUT  DEBUG CONSOLE  TERMINAL                                        1: python          + 回 回 ^ ×
s NT 10.0; Win64; x64) AppleWebKit/537.36 (KHTML, like Gecko) Chrome/91.0.4472.124 Safari/537.36"
phpmyadmin_1  | 172.19.0.1 - - [07/Jul/2021:16:52:14 +0000] "GET /js/vendor/codemirror/addon/hint/show-hint.js?v=5.1.0 HTTP/1.1" 200 5659 "-" "Mozilla/5.0
(Windows NT 10.0; Win64; x64) AppleWebKit/537.36 (KHTML, like Gecko) Chrome/91.0.4472.124 Safari/537.36"
phpmyadmin_1  | 172.19.0.1 - - [07/Jul/2021:16:52:14 +0000] "GET /js/vendor/codemirror/addon/hint/sql-hint.js?v=5.1.0 HTTP/1.1" 200 3097 "-" "Mozilla/5.0 (
Windows NT 10.0; Win64; x64) AppleWebKit/537.36 (KHTML, like Gecko) Chrome/91.0.4472.124 Safari/537.36"
phpmyadmin_1  | 172.19.0.1 - - [07/Jul/2021:16:52:14 +0000] "GET /js/vendor/codemirror/addon/lint/lint.js?v=5.1.0 HTTP/1.1" 200 3069 "-" "Mozilla/5.0 (Wind
ows NT 10.0; Win64; x64) AppleWebKit/537.36 (KHTML, like Gecko) Chrome/91.0.4472.124 Safari/537.36"
phpmyadmin_1  | 172.19.0.1 - - [07/Jul/2021:16:52:14 +0000] "GET /js/dist/codemirror/addon/lint/sql-lint.js?v=5.1.0 HTTP/1.1" 200 810 "-" "Mozilla/5.0 (Win
dows NT 10.0; Win64; x64) AppleWebKit/537.36 (KHTML, like Gecko) Chrome/91.0.4472.124 Safari/537.36"
phpmyadmin_1  | 172.19.0.1 - - [07/Jul/2021:16:52:14 +0000] "GET /js/vendor/codemirror/lib/codemirror.js?v=5.1.0 HTTP/1.1" 200 106200 "-" "Mozilla/5.0 (Win
dows NT 10.0; Win64; x64) AppleWebKit/537.36 (KHTML, like Gecko) Chrome/91.0.4472.124 Safari/537.36"
phpmyadmin_1  | 172.19.0.1 - - [07/Jul/2021:16:52:14 +0000] "GET /js/dist/console.js?v=5.1.0 HTTP/1.1" 200 10628 "-" "Mozilla/5.0 (Windows NT 10.0; Win64;
x64) AppleWebKit/537.36 (KHTML, like Gecko) Chrome/91.0.4472.124 Safari/537.36"
phpmyadmin_1  | 172.19.0.1 - - [07/Jul/2021:16:52:14 +0000] "GET /themes/pmahomme/img/logo_right.png HTTP/1.1" 200 4592 "-" "Mozilla/5.0 (Windows NT 10.0;
Win64; x64) AppleWebKit/537.36 (KHTML, like Gecko) Chrome/91.0.4472.124 Safari/537.36"
phpmyadmin_1  | 172.19.0.1 - - [07/Jul/2021:16:52:14 +0000] "GET /themes/dot.gif HTTP/1.1" 200 325 "-" "Mozilla/5.0 (Windows NT 10.0; Win64; x64) AppleWebK
it/537.36 (KHTML, like Gecko) Chrome/91.0.4472.124 Safari/537.36"
phpmyadmin_1  | 172.19.0.1 - - [07/Jul/2021:16:52:14 +0000] "GET /themes/pmahomme/css/printview.css?v=5.1.0 HTTP/1.1" 200 954 "-" "Mozilla/5.0 (Windows NT
10.0; Win64; x64) AppleWebKit/537.36 (KHTML, like Gecko) Chrome/91.0.4472.124 Safari/537.36"
phpmyadmin_1  | 172.19.0.1 - - [07/Jul/2021:16:52:14 +0000] "GET /themes/pmahomme/img/b_help.png HTTP/1.1" 200 989 "http://localhost:8000/themes/pmahomme/c
ss/theme.css?v=5.1.0&nocache=4842446188ltr&server=1" "Mozilla/5.0 (Windows NT 10.0; Win64; x64) AppleWebKit/537.36 (KHTML, like Gecko) Chrome/91.0.4472.124
Safari/537.36"
phpmyadmin_1  | 172.19.0.1 - - [07/Jul/2021:16:52:14 +0000] "GET /favicon.ico HTTP/1.1" 200 22788 "-" "Mozilla/5.0 (Windows NT 10.0; Win64; x64) AppleWebKi
t/537.36 (KHTML, like Gecko) Chrome/91.0.4472.124 Safari/537.36"
```

這時輸入 ctrl+c，即可離開，但如果想要背景執行 mysql 資料庫，怎麼做呢？執行以下指令，在原先指令上，加入 -d，即可背景執行：

```
docker-compose -f mysql.yml up -d
```

呈現以下畫面：

```
sam@DESKTOP-IKT69L5:~/FinMindBook/DataEngineering/Chapter5/5.3$ docker-compose -f mysql.yml up -d
Creating network "53_dev" with the default driver
Creating 53_mysql_1 ... done
Creating 53_phpmyadmin_1 ... done
sam@DESKTOP-IKT69L5:~/FinMindBook/DataEngineering/Chapter5/5.3$
```

就在背景執行了。

執行以下指令，可以看到目前背景正在執行那些 Docker：

```
docker ps
```

```
sam@DESKTOP-IKT69L5:~/FinMindBook/DataEngineering/Chapter5/5.3$ docker ps
CONTAINER ID   IMAGE                              COMMAND                CREATED         STATUS         PORTS
                                    NAMES
6ab83448cc06   phpmyadmin/phpmyadmin:5.1.0   "/docker-entrypoint..."   14 minutes ago   Up 14 minutes   0.0.0.0:8000->80/tcp, :::800
0->80/tcp                          53_phpmyadmin_1
bc9a660d7186   mysql:8.0                      "docker-entrypoint.s..."   14 minutes ago   Up 14 minutes   0.0.0.0:3306->3306/tcp, :::3
306->3306/tcp, 33060/tcp           53_mysql_1
sam@DESKTOP-IKT69L5:~/FinMindBook/DataEngineering/Chapter5/5.3$ ▌
```

剛剛啟動了 mysql 與 phpmyadmin，那要如何關閉呢？執行以下指令就會
關閉 MySQL：

```
docker-compose -f mysql.yml down
```

那什麼是 MySQL 呢？ MySQL 是一種關聯式資料庫，主要用來儲存結構
型資料，優點是免費，算是蠻廣泛被使用的資料庫，而金融資料，大多
也是結構型資料，因此本書使用 MySQL 進行儲存。

由於金融資料是時間序列型資料，MySQL 是關聯式資料庫，並不完全適
合時序資料，如果想要使用專門的時序資料庫，有以下幾種可做選擇，
InfluxDB 、kdb+、TimescaleDB、DolphinDB 等。

由於用 Docker，以上幾種資料庫都很容易架設，網路上可以找到現成的
Docker Image，像本書架設 MySQL，就是使用現成的 Docker Image。筆
者皆架設後，比較 insert 輸入、select 輸出這兩個最常使用的語法，最後
DolphinDB 勝出，主要是效能上的優勢，當然 DolphinDB 也不是完全沒
有缺點，主要還是看使用情境。如果需求是好上手、通用性，那可以選
擇 MySQL，如果需求是效能，那可以選用專門的時序資料庫。

當然實務上，不會只選擇單一資料庫，使用多種資料庫是可行的。以
FinMind 為例，使用 MySQL、Redis、DolphinDB 三種資料庫，依序處理
結構化資料、即時資料、時間序列資料，根據不同情境，進行調整。

有資料庫之後，可以將上一個章節的三隻爬蟲，抓到的資料，寫進資料
庫儲存，在下個章節，會將介紹上傳資料庫的程式架構。

5.4 上傳資料到資料庫

5.4.1 架構介紹

https://github.com/FinMind/FinMindBook/tree/master/DataEngineering/Chapter5/5.4

```
├── create_partition_table.sql
├── create_table.sql
├── financialdata
│   ├── clients.py
│   ├── __init__.py
│   ├── router.py
│   ├── taifex_crawler.py
│   ├── tpex_crawler.py
│   └── twse_crawler.py
├── financialdata.egg-info
│   ├── dependency_links.txt
│   ├── PKG-INFO
│   ├── SOURCES.txt
│   └── top_level.txt
├── Pipfile
├── Pipfile.lock
├── README.md
└── setup.py
```

架構與 5.2.1 類似,一樣有 Pipfile,不同的是,現在多了 setup.py,這個檔案主要是讓你的專案變成一個 module,可以直接 import。

https://github.com/FinMind/FinMindBook/blob/master/DataEngineering/Chapter5/5.4/setup.py

```python
setup.py

from setuptools import setup, find_packages
from os import path
from io import open


here = path.abspath(path.dirname(__file__))

with open(path.join(here, "README.md"), encoding="utf-8") as f:
    long_description = f.read()

setup(
    name="financialdata",  # Required
    version="1.0.1",  # Required
    description="financial mining",  # Optional
    long_description=long_description,  # Optional
    long_description_content_type="text/markdown",
    # Optional (see note above)
    url="https://github.com/linsamtw",  # Optional
    author="linsam",  # Optional
    author_email="samlin266118@gmail.com",  # Optional
    classifiers=[  # Optional
        "Development Status :: 3 - Alpha",
        "Intended Audience :: Developers",
        "Topic :: Software Development :: Build Tools",
        "License :: OSI Approved :: MIT License",
        "Programming Language :: Python :: 3.6",
    ],
    keywords="financial, python",  # Optional
    project_urls={  # Optional
        "documentation": "https://linsamtw.github.io/FinMindDoc/",
        "Source": "https://github.com/linsamtw/FinMind",
    },
)
```

那要如何將專案變成 module，加入 Pipfile 呢？使用以下指令：

```
pipenv install -e .
```

Pipfile 如下，注意一點，financialdata = {editable = true,path = "."}，就是
將這個專案加入 module。

```
[[source]]
name = "pypi"
url = "https://pypi.org/simple"
verify_ssl = true

[dev-packages]

[packages]
pandas = "*"
requests = "*"
pydantic = "*"
sqlalchemy = "*"
loguru = "*"
financialdata = {editable = true,path = "."}
tqdm = "==4.62.3"

[requires]
python_version = "3.6"
```

除此之外，多了 clients.py、router.py 這兩個檔案，是本章節，要對資料
庫做操作的程式碼，以下將一一介紹。

5.4.2 Clients、Router 資料庫操作

Clients

主要是管理所有對資料庫的連線，將連線有關的程式獨立成一個檔案
管理。以目前來說，雖然只有一個對 MySQL 的連線，但未來不只需
要 MySQL，可能還會使用 Redis、Time Series DB 等，也可能使用多個
MySQL，以 FinMind 為例，有超過 10 個對資料庫的 client，因此，先獨
立出來，對未來的架構設計上，會有很大的幫助。

https://github.com/FinMind/FinMindBook/blob/master/DataEngineering/
Chapter5/5.4/financialdata/clients.py

```
clients.py
from sqlalchemy import create_engine, engine


def get_mysql_financialdata_conn() -> engine.base.Connection:
    """
    user: root
    password: test
    host: localhost
    port: 3306
    database: financialdata
    如果有實體 IP，以上設定可以自行更改
    """
    address = "mysql+pymysql://root:test@localhost:3306/financialdata"
    engine = create_engine(address)
    connect = engine.connect()
    return connect
```

本書選用 SQLAlchemy 套件，主要是能同時支援 ORM（Object-Relational Mapping）用物件方式操作資料庫，跟 Raw SQL 直接寫純 SQL，這兩種方式都有對應的使用場景，不侷限在某一個部分，對未來架構設計上保持彈性。

Router

主要管理對資料庫的 connect 連線、upload 上傳、alive 確認連線是否活著等操作。一般來說，爬蟲與資料庫操作，這兩部分會分開來設計。

https://github.com/FinMind/FinMindBook/blob/master/DataEngineering/Chapter5/5.4/financialdata/router.py

```
router.py
import time
import typing


from loguru import logger
from sqlalchemy import engine


from financialdata import clients


def check_alive(
    connect: engine.base.Connection,
):
    """在每次使用之前，先確認 connect 是否活著"""
    connect.execute("SELECT 1 + 1")
```

```python
def reconnect(
    connect_func: typing.Callable,
) -> engine.base.Connection:
    """如果連線斷掉，重新連線"""
    try:
        connect = connect_func()
    except Exception as e:
        logger.info(
            f"{connect_func.__name__} reconnect error {e}"
        )
    return connect


def check_connect_alive(
    connect: engine.base.Connection,
    connect_func: typing.Callable,
):
    if connect:
        try:
            check_alive(connect)
            return connect
        except Exception as e:
            logger.info(
                f"{connect_func.__name__} connect, error: {e}"
            )
            time.sleep(1)
            connect = reconnect(
                connect_func
            )
            return check_connect_alive(
                connect, connect_func
            )
```

```python
        else:
            connect = reconnect(
                connect_func
            )
            return check_connect_alive(
                connect, connect_func
            )


class Router:
    def __init__(self):
        self._mysql_financialdata_conn = (
            clients.get_mysql_financialdata_conn()
        )

    def check_mysql_financialdata_conn_alive(
        self,
    ):
        self._mysql_financialdata_conn = check_connect_alive(
            self._mysql_financialdata_conn,
            clients.get_mysql_financialdata_conn,
        )
        return (
            self._mysql_financialdata_conn
        )

    @property
    def mysql_financialdata_conn(self):
        """
        使用 property，在每次拿取 connect 時，
        都先經過 check alive 檢查 connect 是否活著
        """
```

```
return (
    self.check_mysql_financialdata_conn_alive()
)
```

5.4.3 建立 table

建立資料表，用來存放爬蟲抓到的資料，以台股股價為例，以下是建立 table 的 SQL，或從以下連結下載：

https://github.com/FinMind/FinMindBook/blob/master/DataEngineering/ Chapter5/5.4/create_table.sql

建立台股股價表：

```sql
create_table.sql

CREATE TABLE `financialdata`.`TaiwanStockPrice`(
    `StockID` VARCHAR(10) NOT NULL,
    `TradeVolume` BIGINT NOT NULL,
    `Transaction` INT NOT NULL,
    `TradeValue` BIGINT NOT NULL,
    `Open` FLOAT NOT NULL,
    `Max` FLOAT NOT NULL,
    `Min` FLOAT NOT NULL,
    `Close` FLOAT NOT NULL,
    `Change` FLOAT NOT NULL,
    `Date` DATE NOT NULL,
    PRIMARY KEY(`StockID`, `Date`)
)
```

複製以上 SQL 語法，貼到以下執行，這樣就會建立 table 了。

同樣的，也建立台股期貨表：

```
CREATE TABLE `financialdata`.`TaiwanFuturesDaily`(
    `Date` DATE NOT NULL,
    `FuturesID` VARCHAR(10) NOT NULL,
    `ContractDate` VARCHAR(30) NOT NULL,
    `Open` FLOAT NOT NULL,
    `Max` FLOAT NOT NULL,
    `Min` FLOAT NOT NULL,
    `Close` FLOAT NOT NULL,
    `Change` FLOAT NOT NULL,
    `ChangePer` FLOAT NOT NULL,
    `Volume` FLOAT NOT NULL,
    `SettlementPrice` FLOAT NOT NULL,
    `OpenInterest` INT NOT NULL,
    `TradingSession` VARCHAR(11) NOT NULL,
    PRIMARY KEY(`FuturesID`, `Date`)
)
```

這時可以到 http://localhost:8000/，會顯示以下畫面，是剛剛建立的 table。

VARCHAR、INT、FLOAT 是資料型態,根據你的資料作調整。設定 PRIMARY KEY 可以保證資料唯一性,也可以增加查詢效率,股價一般都會使用 StockID、Date 作為查詢條件,因此使用以上兩個變數做 PRIMARY KEY,期貨也是同樣邏輯,接下來,將修改程式,從原先存成 csv,改成上傳資料庫。

5.4.4 爬蟲上傳資料庫

相關程式碼如下,可以到以下連結下載:

https://github.com/FinMind/FinMindBook/blob/master/DataEngineering/ Chapter5/5.4/financialdata/twse_crawler.py

```
financialdata/twse_crawler.py
import datetime
import sys
import time
import typing


import pandas as pd
```

```python
import requests
from loguru import logger
from pydantic import BaseModel
from tqdm import tqdm

from financialdata.router import Router


def clear_data(
    df: pd.DataFrame,
) -> pd.DataFrame:
    """資料清理, 將文字轉成數字"""
    df["Dir"] = (
        df["Dir"]
        .str.split(">")
        .str[1]
        .str.split("<")
        .str[0]
    )
    df["Change"] = (
        df["Dir"] + df["Change"]
    )
    df["Change"] = (
        df["Change"]
        .str.replace(" ", "")
        .str.replace("X", "")
        .astype(float)
    )
    df = df.fillna("")
    df = df.drop(["Dir"], axis=1)
    for col in [
        "TradeVolume",
```

```python
        "Transaction",
        "TradeValue",
        "Open",
        "Max",
        "Min",
        "Close",
        "Change",
    ]:
        df[col] = (
            df[col]
            .astype(str)
            .str.replace(",", "")
            .str.replace("X", "")
            .str.replace("+", "")
            .str.replace("----", "0")
            .str.replace("---", "0")
            .str.replace("--", "0")
        )
    return df


def colname_zh2en(
    df: pd.DataFrame,
    colname: typing.List[str],
) -> pd.DataFrame:
    """資料欄位轉換，英文有助於接下來存入資料庫"""
    taiwan_stock_price = {
        "證券代號": "StockID",
        "證券名稱": "",
        "成交股數": "TradeVolume",
        "成交筆數": "Transaction",
        "成交金額": "TradeValue",
```

```python
        "開盤價": "Open",
        "最高價": "Max",
        "最低價": "Min",
        "收盤價": "Close",
        "漲跌(+/-)": "Dir",
        "漲跌價差": "Change",
        "最後揭示買價": "",
        "最後揭示買量": "",
        "最後揭示賣價": "",
        "最後揭示賣量": "",
        "本益比": "",
    }
    df.columns = [
        taiwan_stock_price[col]
        for col in colname
    ]
    df = df.drop([""], axis=1)
    return df


def twse_header():
    """網頁瀏覽時, 所帶的 request header 參數, 模仿瀏覽器發送 request"""
    return {
        "Accept": "application/json, text/javascript, */*; q=0.01",
        "Accept-Encoding": "gzip, deflate",
        "Accept-Language": "zh-TW,zh;q=0.9,en-US;q=0.8,en;q=0.7",
        "Connection": "keep-alive",
        "Host": "www.twse.com.tw",
        "Referer": "https://www.twse.com.tw/zh/page/trading/exchange/
MI_INDEX.html",
        "User-Agent": "Mozilla/5.0 (Windows NT 10.0; Win64;
x64) AppleWebKit/537.36 (KHTML, like Gecko) Chrome/71.0.3578.98
```

```python
Safari/537.36",
        "X-Requested-With": "XMLHttpRequest",
    }


def crawler_twse(
    date: str,
) -> pd.DataFrame:
    """
    證交所網址
    https://www.twse.com.tw/zh/page/trading/exchange/MI_INDEX.html
    """
    # headers 中的 Request url
    url = (
        "https://www.twse.com.tw/exchangeReport/MI_INDEX"
        "?response=json&date={date}&type=ALL"
    )
    url = url.format(
        date=date.replace("-", "")
    )
    # 避免被證交所 ban ip, 在每次爬蟲時, 先 sleep 5 秒
    time.sleep(5)
    # request method
    res = requests.get(
        url, headers=twse_header()
    )
    if (
        res.json()["stat"]
        == "很抱歉，沒有符合條件的資料!"
    ):
        # 如果 date 是周末，會回傳很抱歉，沒有符合條件的資料!
        return pd.DataFrame()
```

```
# 2009 年以後的資料，股價在 response 中的 data9
# 2009 年以後的資料，股價在 response 中的 data8
# 不同格式，在證交所的資料中，是很常見的，
# 沒資料的情境也要考慮進去，例如現在週六沒有交易，但在 2007 年週六是有
交易的
try:
    if "data9" in res.json():
        df = pd.DataFrame(
            res.json()["data9"]
        )
        colname = res.json()[
            "fields9"
        ]
    elif "data8" in res.json():
        df = pd.DataFrame(
            res.json()["data8"]
        )
        colname = res.json()[
            "fields8"
        ]
    elif res.json()["stat"] in [
        "查詢日期小於93年2月11日，請重新查詢!",
        "很抱歉，沒有符合條件的資料!",
    ]:
        return pd.DataFrame()
except BaseException:
    return pd.DataFrame()

if len(df) == 0:
    return pd.DataFrame()
# 欄位中英轉換
df = colname_zh2en(
```

```
        df.copy(), colname
    )
    df["Date"] = date
    return df


class TaiwanStockPrice(BaseModel):
    StockID: str
    TradeVolume: int
    Transaction: int
    TradeValue: int
    Open: float
    Max: float
    Min: float
    Close: float
    Change: float
    Date: str


def check_schema(
    df: pd.DataFrame,
) -> pd.DataFrame:
    """檢查資料型態，確保每次要上傳資料庫前，型態正確"""
    df_dict = df.to_dict("records")
    df_schema = [
        TaiwanStockPrice(**dd).__dict__
        for dd in df_dict
    ]
    df = pd.DataFrame(df_schema)
    return df
```

```python
def gen_date_list(
    start_date: str, end_date: str
) -> typing.List[str]:
    """建立時間列表，用於爬取所有資料"""
    start_date = (
        datetime.datetime.strptime(
            start_date, "%Y-%m-%d"
        ).date()
    )
    end_date = (
        datetime.datetime.strptime(
            end_date, "%Y-%m-%d"
        ).date()
    )
    days = (
        end_date - start_date
    ).days + 1
    date_list = [
        str(
            start_date
            + datetime.timedelta(
                days=day
            )
        )
        for day in range(days)
    ]
    return date_list


def main(
    start_date: str, end_date: str
):
    """證交所寫明，※ 本資訊自民國93年2月11日起提供"""
    date_list = gen_date_list(
```

```
        start_date, end_date
    )
    db_router = Router()
    for date in tqdm(date_list):
        logger.info(date)
        df = crawler_twse(date=date)
        if len(df) > 0:
            # 資料清理
            df = clear_data(df.copy())
            # 檢查資料型態
            df = check_schema(df.copy())
            # upload db
            try:
                df.to_sql(
                    name="TaiwanStockPrice",
                    con=db_router.mysql_financialdata_conn,
                    if_exists="append",
                    index=False,
                    chunksize=1000,
                )
            except Exception as e:
                logger.info(e)


if __name__ == "__main__":
    start_date, end_date = sys.argv[1:]
    main(start_date, end_date)
```

主要使用 dataframe 的 to_sql 指令，上傳資料到資料庫，其他部分沒有做更動。因為在創建 table 時，有設定 PRIMARY KEY，如果同樣的資料已經存在在 table，會有 error，因此使用 try except 做例外處理。

一樣，先安裝 Package。

```
pipenv sync
```

當到新的章節、不同資料夾時，因為使用的 Package 不同，都請先執行以上指令。

執行以下指令，會進行爬蟲，並將資料存到資料庫中。

```
pipenv run python financialdata/twse_crawler.py 2021-01-01 2021-01-10
```

之後切換到 http://localhost:8000/，就會看到爬蟲收集下來的資料了。

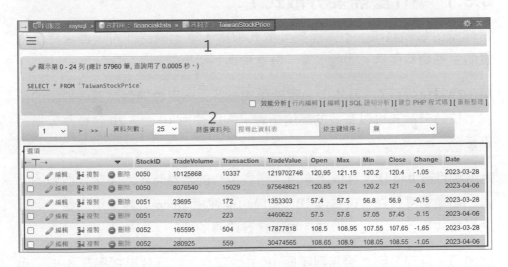

到這裡，成功使用資料庫，管理爬蟲收集到的資料，未來預計所有爬蟲抓到的資料，都用資料庫管理，相對來說也比 csv 好管理。

本書到此，有證交所、櫃買中心、期交所的爬蟲，爬蟲方式大致上都相同，因此爬蟲不是問題，同時也用資料庫管理抓到的資料，預期未來可能需要收集 20、30 種不同資料，這時需要頻繁去證交所、期交所等做爬蟲，可能會被 ban，本書使用 sleep 5 秒作處理，這時會拖累爬蟲效率，下一個章節，將使用分散式，處理效率問題。

一次爬蟲要等 5 秒，那分散在 5 台機器上，平均下來，一次只需等待 1 秒，這就是分散的好處。

5.5　分散式爬蟲

5.5.1　為什麼需要分散式？

當你做少量爬蟲時，不太需要分散式架構，單一機器即可，但目標是大量的金融資料爬蟲，因此，單一機器會遇到以下問題。

單一機器負荷不了。基本上，爬蟲程式一多，一定會遇到需要多機器的情況，這時分散式可以有效的高度拓展，同時使用 10 台、20 台機器分擔爬蟲任務。

過度爬蟲被 ban IP。在做證交所、櫃買中心的爬蟲時，大多數人遇到的問題，都是過度爬蟲，被封鎖 IP，這時可能使用 sleep 去解決這問題，但這就影響效率了，因此分散式，可以提升爬蟲的效能。

例如同一時間，用 5 台機器不同 IP 去做爬蟲，這時效率就提升 5 倍，也不用擔心 sleep 的問題。

單一機器故障。如果單一機器突然故障，那至少還有其他分散式機器可以負荷工作，不至於整個爬蟲掛掉。

5.5.2 分散式任務轉發 --- RabbitMQ

這裡介紹 RabbitMQ 這個任務轉發工具，來進行分散式。在分散式爬蟲架構中，會有三個角色。

Producer，主要發派爬蟲任務給 Broker。

Broker，負責從 Producer 接收任務，並轉發任務給 Worker，這裡使用 RabbitMQ 作為訊息傳遞中心。

Worker，工人，負責從 Broker 接收任務，並且執行任務，以本書為例，就是負責做爬蟲的工作，在架構中，會在不同機器開啟多個 Worker 做爬蟲，達到多 IP、任務分散的效果，因此也不用擔心被證交所 ban IP。

Producer →發送任務→ Broker ←拿取任務← Worker

以下，將使用 Docker 一鍵安裝 RabbitMQ，同時並安裝 Flower 這個 Worker 狀態監控工具。

將以下程式，存成 rabbitmq.yml，或是從以下連結下載：

https://github.com/FinMind/FinMindBook/tree/master/DataEngineering/Chapter5/5.5/5.5.2

```
rabbitmq.yml
version: '3'
services:

  rabbitmq:
    image: 'rabbitmq:3.6-management-alpine'
    ports:
      # docker publish port 5672/15672 to 5672/15672
      # 將 docker 內部 ip 5672/15672, 跟外部 5672/15672 做連結
```

```
    - '5672:5672'
    - '15672:15672'
  environment:
    RABBITMQ_DEFAULT_USER: "worker"
    RABBITMQ_DEFAULT_PASS: "worker"
    RABBITMQ_DEFAULT_VHOST: "/"
  networks:
    - dev

flower:
  image: mher/flower:0.9.5
  command: ["flower", "--broker=amqp://worker:worker@rabbitmq",
"--port=5555"]
  ports:
    # docker publish port 5555 to 5555
    # 將 docker 內部 ip 5555, 跟外部 5555 做連結
    - 5555:5555
  depends_on:
    - rabbitmq
  networks:
    - dev

networks:
  dev:
```

接著，一樣的指令啟動 RabbitMQ：

```
docker-compose -f rabbitmq.yml up -d
```

打開瀏覽器，輸入 http://localhost:15672/，帳號密碼是 worker/worker，
就會看到以下畫面，這就是 RabbitMQ 的畫面：

再打開瀏覽器，輸入 http://localhost:5555，就可以看到 Flower 的畫面，
Flower 是負責監控工人 Worker 的狀態，看 Worker 是否活著，工作是否
正常，目前還沒有啟動 Worker，因此看不到東西。

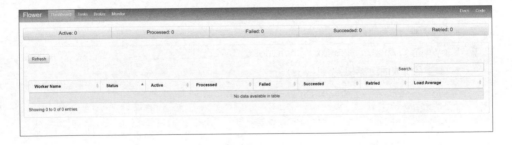

在下個章節，將進入主題，使用爬蟲搭配分散式架構，並且上傳資料到
資料庫。

5.5.3 Python 分散式爬蟲工具 --- Celery

本章節，使用 Celery 這個 Python 的分散式套件，在爬蟲之前，先舉個
Celery 的簡單範例。

整份程式碼，可以從以下連結下載：

https://github.com/FinMind/FinMindBook/tree/master/DataEngineering/
Chapter5/5.5/5.5.3

建立 worker.py，待會會使用這個檔案啟動工人 Worker，Worker 連線
RabbitMQ 之後，只要有任務，工人就會做事。

https://github.com/FinMind/FinMindBook/blob/master/DataEngineering/
Chapter5/5.5/5.5.3/worker.py

```python
worker.py

from celery import Celery


app = Celery(
    "task",
    # 只包含 tasks.py 裡面的程式, 才會成功執行
    include=["tasks"],
    # 連線到 rabbitmq,
    # pyamqp://user:password@localhost:5672/
    # 本書設定的帳號密碼都是 worker
    broker="pyamqp://worker:worker@localhost:5672/",
)
```

建立 tasks.py，註冊 crawler 函數，有註冊的函數，Producer 才能發送。

以下程式印出 crawler、upload db 字串，未來將在這兩個地方，結合 5.4
章節，改成爬蟲與上傳資料庫的程式，目前暫時先用 print 代替。

https://github.com/FinMind/FinMindBook/blob/master/DataEngineering/
Chapter5/5.5/5.5.3/tasks.py

tasks.py

```python
from worker import app
# 註冊 task, 有註冊的 task 才可以變成任務發送給 rabbitmq
@app.task()
def crawler(x):
    print("crawler")
    print("upload db")
    return x
```

這裡寫法比較特別，使用 crawler.delay 發送任務，x=0 是這個函數所需的
參數。

https://github.com/FinMind/FinMindBook/blob/master/DataEngineering/
Chapter5/5.5/5.5.3/producer.py

producer.py

```python
from tasks import crawler

# 發送任務有兩種方式
# 1.
crawler.delay(x=0)
# 2.
# task = crawler.s(x=0)
# task.apply_async()
```

安裝 Package

```
pipenv sync
```

啟動 Celery Worker

```
pipenv run celery -A worker worker --loglevel=info
```

如果執行畫面與下圖相同，代表成功啟動：

```
PROBLEMS   OUTPUT   DEBUG CONSOLE   TERMINAL                              1: python        ∨   +  ⊞  🗑  ∧

sam@DESKTOP-IKT69L5:~/FinMindBook/DataEngineering/Chapter5/5.5/5.5.3$ pipenv run celery -A worker worker --loglevel=info

 -------------- celery@DESKTOP-IKT69L5 v5.0.5 (singularity)
--- ***** -----
-- ******* ---- Linux-4.4.0-19041-Microsoft-x86_64-with-Ubuntu-18.04-bionic 2021-10-30 17:03:01
- *** --- * ---
- ** ---------- [config]
- ** ---------- .> app:         task:0x7f61279fe4e0
- ** ---------- .> transport:   amqp://worker:**@localhost:5672//
- ** ---------- .> results:     disabled://
- *** --- * --- .> concurrency: 12 (prefork)
-- ******* ---- .> task events: OFF (enable -E to monitor tasks in this worker)
--- ***** -----
 -------------- [queues]
                .> celery           exchange=celery(direct) key=celery

[tasks]
  . tasks.crawler

[2021-10-30 17:03:01,641: INFO/MainProcess] Connected to amqp://worker:**@127.0.0.1:5672//
[2021-10-30 17:03:01,668: INFO/MainProcess] mingle: searching for neighbors
[2021-10-30 17:03:02,713: INFO/MainProcess] mingle: all alone
[2021-10-30 17:03:02,760: INFO/MainProcess] celery@DESKTOP-IKT69L5 ready.
[2021-10-30 17:03:05,830: INFO/MainProcess] Events of group {task} enabled by remote.
```

打開 flower http://localhost:5555/ 畫面，就會看到剛剛的 Worker 的狀態，
Status 是 Online。

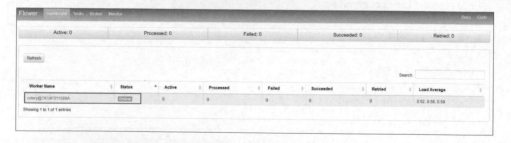

如果關閉 Celery，Status 就會變成 Offline，這就是 Flower 的用途，用來
監控工人的狀態。

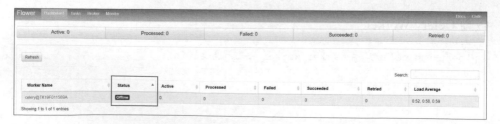

接下來，使用以下指令，發送任務。

```
pipenv run python producer.py
```

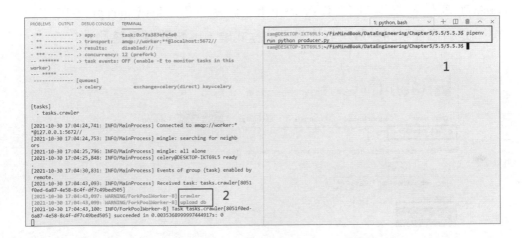

Worker 接到任務之後，就會開始做事，印出 crawler、upload db 這兩個字串，到這邊，已經成功使用 RabbitMQ 分散式架構了。

那 RabbitMQ 用在哪呢？這時先把 Worker 關掉，並且再發一次任務，並打開 http://localhost:15672/

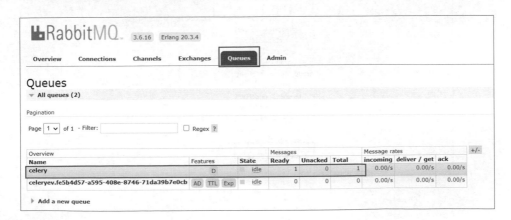

點選 **queue**，底下出現 Celery 的 Name，這就是剛剛發送的任務，再點選 **Celery**，跳轉頁面後，在 **Get messages** 部分，點選 **Get Message(s)** 的按鈕，可以看到這個 **task** 的訊息，特別是 **task** 與 **kwargsrepr** 這兩部分，**task** 顯示任務是要執行 **Tasks.crawler** 這個函數，**kwargsrepr** 中的 {'x': 0}，是這個任務的參數，對比上面的程式碼，應該就很清楚了。

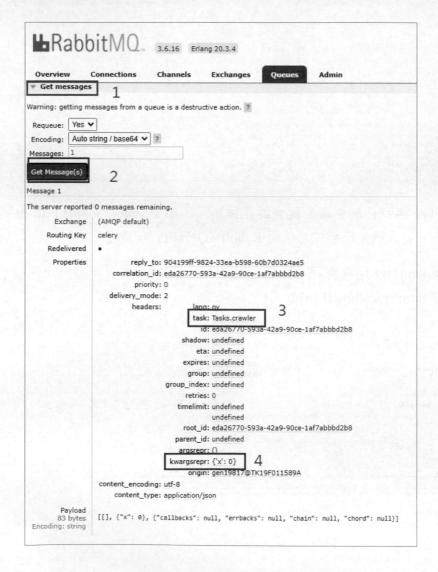

本章節用個簡單例子展示 RabbitMQ、Celery 這個工具，讀者只要將印出 crawler、upload db 的這兩部分，改成爬蟲、上傳資料庫的程式碼，就可以順利達到分散式爬蟲的效果。

5.5.4 真實應用場景

在 5.5.3 中，Worker、RabbitMQ 都是在同一台機器上，真正要達到分散的場景，至少要準備兩台機器，這時一般會遇到固定 IP 和機器不夠的問題，可以參考 Chapter4 中的 Linode 介紹，一個月只要 5 美金，開兩台機器一個月只要 10 美金，約 300 台幣，比單純租網路還便宜，以下示範多 ip 的設定。

請先到 Linode 申請 2 組 5 美金的機器，Ubuntu 系統，這時你會得到 2 組 IP，本書用 rabbitmq_ip、worker1_ip 代替。

設定 Linode 步驟：

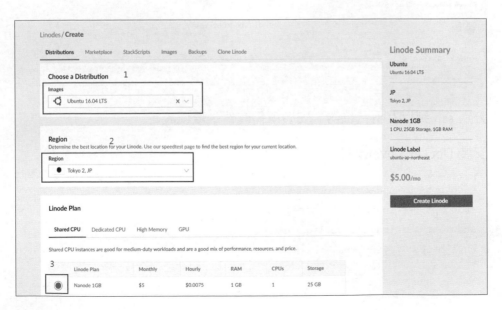

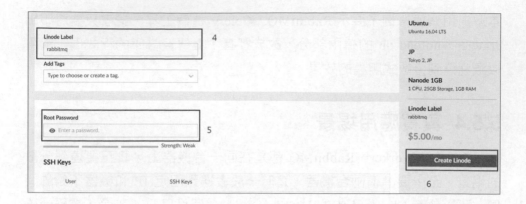

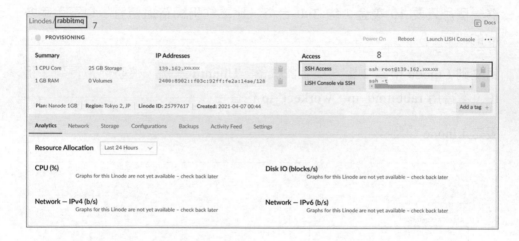

分別 ssh 到以上兩台機器，安裝所需套件、環境設定。

安裝 Docker、pyenv：

```
sudo apt-get update
sudo apt-get install -y docker.io
sudo apt-get install curl git bzip2 -y
curl https://pyenv.run | bash
```

環境設定：

```
echo 'export LC_ALL=C.UTF-8' >> ~/.bashrc
echo 'export LANG=C.UTF-8' >> ~/.bashrc
echo 'export PYENV_ROOT="$HOME/.pyenv"' >> ~/.bashrc
echo 'export PATH="$PYENV_ROOT/shims:$PATH"' >> ~/.bashrc
echo 'export PATH="$PYENV_ROOT/bin:$PATH"' >> ~/.bashrc
echo -e 'if command -v pyenv 1>/dev/null 2>&1; then\n  eval "$(pyenv
init -)"\nfi' >> ~/.bashrc
exec $SHELL
```

安裝 miniconda3-4.3.30：

```
pyenv install miniconda3-4.3.30
pyenv global miniconda3-4.3.30
```

clone 程式碼、安裝套件：

```
pip install docker-compose pipenv
git clone https://github.com/FinMind/FinMindBook.git
cd /FinMindBook/DataEngineering/Chapter5/5.5/5.5.4/
pipenv sync
```

到 rabbitmq_ip 的機器，架設 RabbitMQ：

```
cd /FinMindBook/DataEngineering/Chapter5/5.5/5.5.4/
docker-compose -f rabbitmq.yml up -d
```

這時瀏覽 http://rabbitmq_ip:15672/，會看到以下畫面：

再到 worker1_ip 的機器，啟動 Celery

```
cd /FinMindBook/DataEngineering/Chapter5/5.5/5.5.4/
pipenv run celery -A worker worker --loglevel=info
```

這時到 http://rabbitmq_ip:5555/ 會看到以下畫面

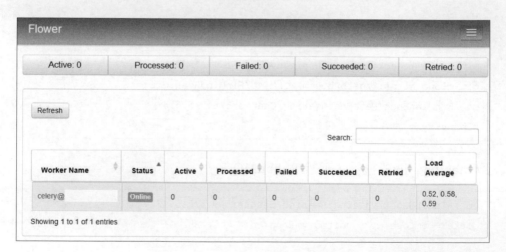

發送任務

```
pipenv run python producer.py
```

成功執行

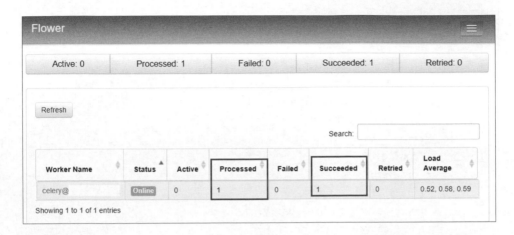

這時就達到分散式了,你可以再多開幾台 Worker,讓 Worker 去處理你的爬蟲任務,並使用 5.3 教學,在雲端架設資料庫,將爬蟲抓到的資料,存在資料庫。

本書到此,已經能成功在雲端上使用分散式爬蟲,並能高度拓展,而資料庫也成功架設了,只要持續新增爬蟲,就能簡單的收集各種資料,下一個章節,將會擷取部分 FinMind 架構做介紹,統整爬蟲、資料庫、分散式。

5.5.5 FinMind 分散式爬蟲架構

可以到以下連結下載程式碼:

https://github.com/FinMind/FinMindBook/tree/master/DataEngineering/Chapter5/5.5/5.5.5

擷取部分 FinMind 爬蟲架構來進行展示。

```
├──── create_partition_table.sql
├──── financialdata
│    ├──── backend
│    │    ├──── db
│    │    │    ├──── clients.py
│    │    │    ├──── db.py
│    │    │    ├──── __init__.py
│    │    │    └──── router.py
│    │    └──── __init__.py
│    ├──── config.py
│    ├──── crawler
│    │    ├──── __init__.py
│    │    ├──── taiwan_futures_daily.py
│    │    └──── taiwan_stock_price.py
│    ├──── __init__.py
│    ├──── producer.py
│    ├──── schema
│    │    ├──── dataset.py
│    │    ├──── __init__.py
│    └──── tasks
│         ├──── __init__.py
│         ├──── task.py
│         └──── worker.py
├──── genenv.py
├──── local.ini
├──── Makefile
├──── mysql.yml
├──── Pipfile
├──── Pipfile.lock
├──── rabbitmq.yml
├──── README.md
└──── setup.py
```

mysql.yml、rabbitmq.yml 分別架設相對應的服務，其餘架構主要分成幾個部分，**backend**、**crawler**、**schema**、**tasks**、**config.py**、**producer.py**，**backend**

資料夾裡，主要是 5.4.2 提到過的 clients.py、db.py、router.py，管理對資料庫的操作：

https://github.com/FinMind/FinMindBook/blob/master/DataEngineering/
Chapter5/5.5/5.5.5/financialdata/backend/db/clients.py

clients.py，建立對資料庫的連線。

```python
from financialdata.config import (
    MYSQL_DATA_USER,
    MYSQL_DATA_PASSWORD,
    MYSQL_DATA_HOST,
    MYSQL_DATA_PORT,
    MYSQL_DATA_DATABASE,
)
from sqlalchemy import create_engine, engine


def get_mysql_financialdata_conn() -> engine.base.Connection:
    address = (
        f"mysql+pymysql://{MYSQL_DATA_USER}:{MYSQL_DATA_PASSWORD}"
        f"@{MYSQL_DATA_HOST}:{MYSQL_DATA_PORT}/{MYSQL_DATA_DATABASE}"
    )
    engine = create_engine(address)
    connect = engine.connect()
    return connect
```

https://github.com/FinMind/FinMindBook/blob/master/DataEngineering/
Chapter5/5.5/5.5.5/financialdata/backend/db/db.py

db.py 對資料庫操作，其中要注意的是，這裡使用兩種方式上傳資料到資料庫：

1. 用 pandas 原生方式上傳
2. 用 SQL 語法上傳

pandas 的方式上傳速度較快，但遇到重複資料，而 table 又有設主鍵時，會發生錯誤，這時需要寫 SQL 語法才能正確上傳，主要是 SQL 語法中特別加入一段，DUPLICATE KEY UPDATE，可以在遇到重複資料時，改成 update 更新的方式上傳資料，藉此避免因為重複資料，導致資料上傳失敗的問題。

```python
import typing

import pandas as pd
import pymysql
from loguru import logger
from sqlalchemy import engine


def update2mysql_by_pandas(
    df: pd.DataFrame,
    table: str,
    mysql_conn: engine.base.Connection,
):
    if len(df) > 0:
        try:
            df.to_sql(
                name=table,
                con=mysql_conn,
                if_exists="append",
                index=False,
                chunksize=1000,
            )
        except Exception as e:
```

```python
        pass


def build_update_sql(
    colname: typing.List[str],
    value: typing.List[str],
):
    update_sql = ",".join(
        [
            ' `{}` = "{}" '.format(
                colname[i],
                str(value[i]),
            )
            for i in range(len(colname))
            if str(value[i])
        ]
    )
    return update_sql


def build_df_update_sql(
    table: str, df: pd.DataFrame
) -> typing.List[str]:
    logger.info("build_df_update_sql")
    df_columns = list(df.columns)
    sql_list = []
    for i in range(len(df)):
        temp = list(df.iloc[i])
        value = [
            pymysql.converters.escape_string(
                str(v)
            )
```

```python
            for v in temp
        ]
        sub_df_columns = [
            df_columns[j]
            for j in range(len(temp))
        ]
        update_sql = build_update_sql(
            sub_df_columns, value
        )
        # SQL 上傳資料方式
        # DUPLICATE KEY UPDATE 意思是
        # 如果有重複，就改用 update 的方式
        # 避免重複上傳
        sql = """INSERT INTO `{}`({})VALUES ({}) ON DUPLICATE KEY UPDATE {}
            """.format(
            table,
            "`{}`".format(
                "`,`".join(
                    sub_df_columns
                )
            ),
            '"{}"'.format(
                '","'.join(value)
            ),
            update_sql,
        )
        sql_list.append(sql)
    return sql_list


def update2mysql_by_sql(
    df: pd.DataFrame,
```

```
    table: str,
    mysql_conn: engine.base.Connection,
):
    sql = build_df_update_sql(table, df)
    commit(
        sql=sql, mysql_conn=mysql_conn
    )


def commit(
    sql: typing.Union[
        str, typing.List[str]
    ],
    mysql_conn: engine.base.Connection = None,
):
    logger.info("commit")
    try:
        trans = mysql_conn.begin()
        if isinstance(sql, list):
            for s in sql:
                try:
                    mysql_conn.execution_options(
                        autocommit=False
                    ).execute(
                        s
                    )
                except Exception as e:
                    logger.info(e)
                    logger.info(s)
                    break

        elif isinstance(sql, str):
```

```
            mysql_conn.execution_options(
                autocommit=False
            ).execute(
                sql
            )
        trans.commit()
    except Exception as e:
        trans.rollback()
        logger.info(e)


def upload_data(
    df: pd.DataFrame,
    table: str,
    mysql_conn: engine.base.Connection,
):
    if len(df) > 0:
        # 直接上傳
        if update2mysql_by_pandas(
            df=df,
            table=table,
            mysql_conn=mysql_conn,
        ):
            pass
        else:
            # 如果有重複的資料
            # 使用 SQL 語法上傳資料
            update2mysql_by_sql(
                df=df,
                table=table,
                mysql_conn=mysql_conn,
            )
```

https://github.com/FinMind/FinMindBook/blob/master/DataEngineering/
Chapter5/5.5/5.5.5/financialdata/backend/db/router.py

router.py 管理 client，並檢查 connect 連線是否活著。

```python
import time
import typing

from loguru import logger
from sqlalchemy import engine
from financialdata.backend.db import clients


def check_alive(connect: engine.base.Connection):
    connect.execute("SELECT 1 + 1")


def check_connect_alive(
    connect: engine.base.Connection,
    connect_func: typing.Callable,
):
    if connect:
        try:
            check_alive(connect)
            return connect
        except Exception as e:
            logger.info(
                f"""
                {connect_func.__name__} reconnect, error: {e}
                """
            )
            time.sleep(1)
```

```
        try:
            connect = connect_func()
        except Exception as e:
            logger.info(
                f"""
                {connect_func.__name__} connect error, error: {e}
                """
            )
        return check_connect_alive(connect, connect_func)


class Router:
    def __init__(self):
        self._mysql_financialdata_conn = clients.get_mysql_financialdata_
conn()

    def check_mysql_financialdata_conn_alive(self):
        self._mysql_financialdata_conn = check_connect_alive(
            self._mysql_financialdata_conn,
            clients.get_mysql_financialdata_conn,
        )
        return self._mysql_financialdata_conn

    @property
    def mysql_financialdata_conn(self):
        return self.check_mysql_financialdata_conn_alive()

    def close_connection(self):
        self._mysql_financialdata_conn.close()
```

crawler 資料夾中，是 taiwan_stock_price.py 和 taiwan_futures_daily.py，
台股股價爬蟲、台股期貨爬蟲，大致上與 5.4 程式的相同，主要是合併證

交所爬蟲 twse_crawler.py 與櫃買中心爬蟲 tpex_crawler.py，統整為台股股價爬蟲 taiwan_stock_price.py，程式如下：

https://github.com/FinMind/FinMindBook/blob/master/DataEngineering/Chapter5/5.5/5.5.5/financialdata/crawler/taiwan_stock_price.py

```
taiwan_stock_price.py
import datetime
import time
import typing

import pandas as pd
import requests
from loguru import logger
from financialdata.schema.dataset import (
    check_schema,
)


def is_weekend(day: int) -> bool:
    return day in [0, 6]


def gen_task_paramter_list(
    start_date: str, end_date: str
) -> typing.List[str]:
    start_date = (
        datetime.datetime.strptime(
            start_date, "%Y-%m-%d"
        ).date()
    )
    end_date = (
        datetime.datetime.strptime(
```

```
            end_date, "%Y-%m-%d"
        ).date()
    )
    days = (
        end_date - start_date
    ).days + 1
    date_list = [
        start_date
        + datetime.timedelta(days=day)
        for day in range(days)
    ]
    # 排除掉周末非交易日
    date_list = [
        dict(
            date=str(d),
            data_source=data_source,
        )
        for d in date_list
        for data_source in [
            "twse",
            "tpex",
        ]
        if not is_weekend(d.weekday())
    ]
    return date_list



def clear_data(
    df: pd.DataFrame,
) -> pd.DataFrame:
    """資料清理，將文字轉成數字"""
    for col in [
```

```python
        "TradeVolume",
        "Transaction",
        "TradeValue",
        "Open",
        "Max",
        "Min",
        "Close",
        "Change",
    ]:
        df[col] = (
            df[col]
            .astype(str)
            .str.replace(",", "")
            .str.replace("X", "")
            .str.replace("+", "")
            .str.replace("----", "0")
            .str.replace("---", "0")
            .str.replace("--", "0")
            .str.replace(" ", "")
            .str.replace("除權息", "0")
            .str.replace("除息", "0")
            .str.replace("除權", "0")
        )
    return df


def colname_zh2en(
    df: pd.DataFrame,
    colname: typing.List[str],
) -> pd.DataFrame:
    """資料欄位轉換，英文有助於接下來存入資料庫"""
    taiwan_stock_price = {
```

```python
        "證券代號": "StockID",
        "證券名稱": "",
        "成交股數": "TradeVolume",
        "成交筆數": "Transaction",
        "成交金額": "TradeValue",
        "開盤價": "Open",
        "最高價": "Max",
        "最低價": "Min",
        "收盤價": "Close",
        "漲跌(+/-)": "Dir",
        "漲跌價差": "Change",
        "最後揭示買價": "",
        "最後揭示買量": "",
        "最後揭示賣價": "",
        "最後揭示賣量": "",
        "本益比": "",
    }
    df.columns = [
        taiwan_stock_price[col]
        for col in colname
    ]
    df = df.drop([""], axis=1)
    return df


def twse_header():
    """網頁瀏覽時, 所帶的 request header 參數, 模仿瀏覽器發送 request"""
    return {
        "Accept": "application/json, text/javascript, */*; q=0.01",
        "Accept-Encoding": "gzip, deflate",
        "Accept-Language": "zh-TW,zh;q=0.9,en-US;q=0.8,en;q=0.7",
        "Connection": "keep-alive",
```

```
        "Host": "www.twse.com.tw",
        "Referer": "https://www.twse.com.tw/zh/page/trading/exchange/
MI_INDEX.html",
        "User-Agent": "Mozilla/5.0 (Windows NT 10.0; Win64;
x64) AppleWebKit/537.36 (KHTML, like Gecko) Chrome/71.0.3578.98
Safari/537.36",
        "X-Requested-With": "XMLHttpRequest",
    }


def tpex_header():
    """網頁瀏覽時, 所帶的 request header 參數, 模仿瀏覽器發送 request"""
    return {
        "Accept": "application/json, text/javascript, */*; q=0.01",
        "Accept-Encoding": "gzip, deflate",
        "Accept-Language": "zh-TW,zh;q=0.9,en-US;q=0.8,en;q=0.7",
        "Connection": "keep-alive",
        "Host": "www.tpex.org.tw",
        "Referer": "https://www.tpex.org.tw/web/stock/aftertrading/
otc_quotes_no1430/stk_wn1430.php?l=zh-tw",
        "User-Agent": "Mozilla/5.0 (Windows NT 10.0; Win64; x64)
AppleWebKit/537.36 (KHTML, like Gecko) Chrome/73.0.3683.103
Safari/537.36",
        "X-Requested-With": "XMLHttpRequest",
    }


def set_column(
    df: pd.DataFrame,
) -> pd.DataFrame:
    """設定資料欄位名稱"""
    df.columns = [
```

```
        "StockID",
        "Close",
        "Change",
        "Open",
        "Max",
        "Min",
        "TradeVolume",
        "TradeValue",
        "Transaction",
    ]
    return df


def crawler_tpex(
    date: str,
) -> pd.DataFrame:
    """
    櫃買中心網址
    https://www.tpex.org.tw/web/stock/aftertrading/otc_quotes_no1430/
stk_wn1430.php?l=zh-tw
    """
    logger.info("crawler_tpex")
    # headers 中的 Request url
    url = "https://www.tpex.org.tw/web/stock/aftertrading/otc_quotes_
no1430/stk_wn1430_result.php?l=zh-tw&d={date}&se=AL"
    url = url.format(
        date=convert_date(date)
    )
    # 避免被櫃買中心 ban ip, 在每次爬蟲時, 先 sleep 5 秒
    time.sleep(5)
    # request method
    res = requests.get(
```

```
        url, headers=tpex_header()
    )
    data = res.json().get("aaData", [])
    df = pd.DataFrame(data)
    if not data or len(df) == 0:
        return pd.DataFrame()
    # 櫃買中心回傳的資料, 並無資料欄位, 因此這裡直接用 index 取特定欄位
    df = df[[0, 2, 3, 4, 5, 6, 7, 8, 9]]
    # 欄位中英轉換
    df = set_column(df.copy())
    df["Date"] = date
    df = clear_data(df.copy())
    return df


def crawler_twse(
    date: str,
) -> pd.DataFrame:
    """
    證交所網址
    https://www.twse.com.tw/zh/page/trading/exchange/MI_INDEX.html
    """
    logger.info("crawler_twse")
    # headers 中的 Request url
    url = "https://www.twse.com.tw/exchangeReport/MI_INDEX?response=
json&date={date}&type=ALL"
    url = url.format(
        date=date.replace("-", "")
    )
    # 避免被證交所 ban ip, 在每次爬蟲時, 先 sleep 5 秒
    time.sleep(5)
    # request method
```

```python
res = requests.get(
    url, headers=twse_header()
)
# 2009 年以後的資料，股價在 response 中的 data9
# 2009 年以後的資料，股價在 response 中的 data8
# 不同格式，在證交所的資料中，是很常見的，
# 沒資料的情境也要考慮進去，例如現在週六沒有交易，但在 2007 年週六是有
交易的
df = pd.DataFrame()
try:
    if "data9" in res.json():
        df = pd.DataFrame(
            res.json()["data9"]
        )
        colname = res.json()[
            "fields9"
        ]
    elif "data8" in res.json():
        df = pd.DataFrame(
            res.json()["data8"]
        )
        colname = res.json()[
            "fields8"
        ]
    elif res.json()["stat"] in [
        "查詢日期小於93年2月11日，請重新查詢!",
        "很抱歉，沒有符合條件的資料!",
    ]:
        pass
except Exception as e:
    logger.error(e)
    return pd.DataFrame()
```

```python
    if len(df) == 0:
        return pd.DataFrame()
    # 欄位中英轉換
    df = colname_zh2en(
        df.copy(), colname
    )
    df["Date"] = date
    df = convert_change(df.copy())
    df = clear_data(df.copy())
    return df


def convert_change(
    df: pd.DataFrame,
) -> pd.DataFrame:
    logger.info("convert_change")
    df["Dir"] = (
        df["Dir"]
        .str.split(">")
        .str[1]
        .str.split("<")
        .str[0]
    )
    df["Change"] = (
        df["Dir"] + df["Change"]
    )
    df["Change"] = (
        df["Change"]
        .str.replace(" ", "")
        .str.replace("X", "")
        .astype(float)
```

```
    )
    df = df.fillna("")
    df = df.drop(["Dir"], axis=1)
    return df


def convert_date(date: str) -> str:
    logger.info("convert_date")
    year, month, day = date.split("-")
    year = int(year) - 1911
    return f"{year}/{month}/{day}"


def crawler(
    parameter: typing.Dict[
        str,
        typing.List[
            typing.Union[
                str, int, float
            ]
        ],
    ]
) -> pd.DataFrame:
    logger.info(parameter)
    date = parameter.get("date", "")
    data_source = parameter.get(
        "data_source", ""
    )
    if data_source == "twse":
        df = crawler_twse(date)
    elif data_source == "tpex":
        df = crawler_tpex(date)
```

```
    df = check_schema(
        df.copy(),
        dataset="TaiwanStockPrice",
    )
    return df
```

期貨爬蟲只是單純將 taifex_crawler.py 改名為 taiwan_futures_daily.py，程式碼與 5.4 相同，在這就不多做介紹。

schema 資料夾中，是統整所有爬蟲的 schema，統整在一起比較好管理。

dataset.py

```
from pydantic import BaseModel
import importlib

import pandas as pd


class TaiwanStockPrice(BaseModel):
    StockID: str
    TradeVolume: int
    Transaction: int
    TradeValue: int
    Open: float
    Max: float
    Min: float
    Close: float
    Change: float
    Date: str


class TaiwanFuturesDaily(BaseModel):
    Date: str
```

```
    FuturesID: str
    ContractDate: str
    Open: float
    Max: float
    Min: float
    Close: float
    Change: float
    ChangePer: float
    Volume: float
    SettlementPrice: float
    OpenInterest: int
    TradingSession: str


def check_schema(df: pd.DataFrame, dataset: str) -> pd.DataFrame:
    """檢查資料型態，確保每次要上傳資料庫前，型態正確"""
    df_dict = df.to_dict("records")
    schema = getattr(
        importlib.import_module("financialdata.schema.dataset"),
        dataset,
    )
    df_schema = [schema(**dd).__dict__ for dd in df_dict]
    df = pd.DataFrame(df_schema)
    return df
```

tasks 資料夾中，是整理 5.5.3 的分散式架構，包含 worker.py、task.py。

worker.py

```
from celery import Celery
from financialdata.config import (
    WORKER_ACCOUNT,
    WORKER_PASSWORD,
    MESSAGE_QUEUE_HOST,
```

```
    MESSAGE_QUEUE_PORT,
)

broker = (
    f"pyamqp://{WORKER_ACCOUNT}:{WORKER_PASSWORD}@"
    f"{MESSAGE_QUEUE_HOST}:{MESSAGE_QUEUE_PORT}/"
)
app = Celery(
    "task",
    include=["financialdata.tasks.task"],
    broker=broker,
)
```

```
task.py
```
```
import importlib
import typing

from financialdata.backend import db
from financialdata.tasks.worker import (
    app,
)


# 註冊 task, 有註冊的 task 才可以變成任務發送給 rabbitmq
@app.task()
def crawler(
    dataset: str,
    parameter: typing.Dict[str, str],
):
    # 使用 getattr, importlib,
    # 根據不同 dataset, 使用相對應的 crawler 收集資料
    # 爬蟲
```

```
    df = getattr(
        importlib.import_module(
            f"financialdata.crawler.{dataset}"
        ),
        "crawler",
    )(parameter=parameter)
    # 上傳資料庫
    db_dataset = dict(
        taiwan_stock_price="TaiwanStockPrice",
        taiwan_futures_daily="TaiwanFuturesDaily",
    )
    db.upload_data(
        df,
        db_dataset,
        db.router.mysql_financialdata_conn,
    )
```

producer.py 跟 5.5.3 一樣，是負責發送任務的，這裡預期可以針對不同爬蟲，統一一個規格去發送任務，並將爬蟲任務根據不同日期，切成一個個獨立的爬蟲，才方便使用分散式去獨力完成，因此使用以下設計。

```
producer.py

import importlib
import sys


from loguru import logger


from financialdata.backend import db
from financialdata.tasks.task import crawler



def Update(dataset: str, start_date: str, end_date: str):
```

```
# 拿取每個爬蟲任務的參數列表,
# 包含爬蟲資料的日期 date,例如 2021-04-10 的台股股價,
# 資料來源 data_source,例如 twse 證交所、tpex 櫃買中心
parameter_list = getattr(
    importlib.import_module(f"financialdata.crawler.{dataset}"),
    "gen_task_paramter_list",
)(start_date=start_date, end_date=end_date)
# 用 for loop 發送任務
for parameter in parameter_list:
    logger.info(f"{dataset}, {parameter}")
    task = crawler.s(dataset, parameter)
    # queue 參數,可以指定要發送到特定 queue 列隊中
    task.apply_async(queue=parameter.get("data_source", ""))


    db.router.close_connection()



if __name__ == "__main__":
    dataset, start_date, end_date = sys.argv[1:]
    Update(dataset, start_date, end_date)
```

統一使用 **gen_task_paramter_list** 產生每個任務的參數列表,並統一使用 crawler 做爬蟲回傳 dataframe,因此規範每隻爬蟲必須要有 **gen_task_paramter_list**、**crawler** 這兩個 function。

config.py 是整份程式用到的設定檔,包含對**資料庫**、**RabbitMQ** 的設定,那為什麼要特別抽出來呢?

```
config.py

import os


MYSQL_DATA_HOST = os.environ.get("MYSQL_DATA_HOST", "127.0.0.1")
```

```
MYSQL_DATA_USER = os.environ.get("MYSQL_DATA_USER", "root")
MYSQL_DATA_PASSWORD = os.environ.get("MYSQL_DATA_PASSWORD", "test")
MYSQL_DATA_PORT = int(os.environ.get("MYSQL_DATA_PORT", "3306"))
MYSQL_DATA_DATABASE = os.environ.get("MYSQL_DATA_DATABASE", "financialdata")

WORKER_ACCOUNT = os.environ.get("WORKER_ACCOUNT", "worker")
WORKER_PASSWORD = os.environ.get("WORKER_PASSWORD", "worker")

MESSAGE_QUEUE_HOST = os.environ.get("MESSAGE_QUEUE_HOST", "127.0.0.1")
MESSAGE_QUEUE_PORT = int(os.environ.get("MESSAGE_QUEUE_PORT", "5672"))
```

第一,是為了方便管理。

第二,在開發與實際產品部屬上,會使用不同設定,開發 dev 就用 localhost,實際部屬 release 就用雲端 IP,甚至在上正式產品環境時,還會經過一層 staging 測試機的環境,因此把環境變數抽出來,有助於未來的開發上。

那環境變數,一般怎麼設定呢?

這裡 FinMind 使用 local.ini 管理開發、測試、正式環境變數:

```ini
local.ini
# dev 環境
[DEFAULT]
MYSQL_DATA_HOST = 127.0.0.1
MYSQL_DATA_USER = root
MYSQL_DATA_PASSWORD = test
MYSQL_DATA_PORT = 3306
MYSQL_DATA_DATABASE = financialdata
WORKER_ACCOUNT = worker
WORKER_PASSWORD = worker
MESSAGE_QUEUE_HOST = 127.0.0.1
```

```
MESSAGE_QUEUE_PORT = 5672

# 測試站環境，這邊可以換成自己測試機的 IP，如果沒有測試機，可以先略過這段
[STAGING]
MYSQL_DATA_HOST = 127.0.0.1
MYSQL_DATA_USER = root
MYSQL_DATA_PASSWORD = test
MYSQL_DATA_PORT = 3306
MYSQL_DATA_DATABASE = financialdata
WORKER_ACCOUNT = worker
WORKER_PASSWORD = worker
MESSAGE_QUEUE_HOST = 127.0.0.1
MESSAGE_QUEUE_PORT = 5672

# 正式站環境，這邊可以換成 linode 上的 IP
[RELEASE]
MYSQL_DATA_HOST = 127.0.0.1
MYSQL_DATA_USER = root
MYSQL_DATA_PASSWORD = test
MYSQL_DATA_PORT = 3306
MYSQL_DATA_DATABASE = financialdata
WORKER_ACCOUNT = worker
WORKER_PASSWORD = worker
MESSAGE_QUEUE_HOST = 127.0.0.1
MESSAGE_QUEUE_PORT = 5672
```

使用 genenv.py 建立 .env 環境檔，搭配 pipenv 使用，pipenv run python
會同時使用 .env 做為環境變數，並且執行 Python，這就不需要額外設定
機器的環境變數：

genenv.py

```
import socket
```

```python
import os
from configparser import ConfigParser

HOME_PATH = "/".join(os.path.abspath(__file__).split("/")[:-1])
HOST_NAME = socket.gethostname()

local_config = ConfigParser()
local_config.read("local.ini")
if os.environ.get("VERSION", ""):
    section = local_config[os.environ.get("VERSION", "")]
elif HOST_NAME in local_config:
    section = local_config[HOST_NAME]
else:
    section = local_config["DEFAULT"]

env_content = ""
for sec in section:
    env_content += "{}={}\n".format(sec.upper(), section[sec])

with open(".env", "w", encoding="utf8") as env:
    env.write(env_content)
```

而新增的 Makefile 是協助紀錄一些常用到的指令，包含啟動資料庫、
RabbitMQ、Celery、發送任務、建立環境變數等。

```makefile
Makefile

# 啟動 mysql
create-mysql:
    docker-compose -f mysql.yml up -d

# 啟動 rabbitmq
create-rabbitmq:
```

```
    docker-compose -f rabbitmq.yml up -d

# 安裝環境
install-python-env:
    pipenv sync

# 啟動 celery, 專門執行 twse queue 列隊的任務,
run-celery-twse:
    pipenv run celery -A financialdata.tasks.worker worker
--loglevel=info --concurrency=1  --hostname=%h -Q twse

# 啟動 celery, 專門執行 tpex queue 列隊的任務
run-celery-tpex:
    pipenv run celery -A financialdata.tasks.worker worker
--loglevel=info --concurrency=1  --hostname=%h -Q tpex

# sent task
sent-taiwan-stock-price-task:
    pipenv run python financialdata/producer.py taiwan_stock_price 2021-
04-01 2021-04-12

# 建立 dev 環境變數
gen-dev-env-variable:
    python genenv.py

# 建立 staging 環境變數
gen-staging-env-variable:
    VERSION=STAGING python genenv.py

# 建立 release 環境變數
gen-release-env-variable:
    VERSION=RELEASE python genenv.py
```

安裝 make

```
sudo apt install make
```

使用 make 建立環境變數

```
make gen-dev-env-variable
```

這時就會產生 .env 的檔案，也就是上面提到的。

只需要簡單一個指令，就能任意切換開發、測試、正式環境，是不是很簡單呢？這就是將 config 抽出來的好處：

```
.env
MYSQL_DATA_HOST=127.0.0.1
MYSQL_DATA_USER=root
MYSQL_DATA_PASSWORD=test
MYSQL_DATA_PORT=3306
MYSQL_DATA_DATABASE=financialdata
WORKER_ACCOUNT=worker
WORKER_PASSWORD=worker
MESSAGE_QUEUE_HOST=127.0.0.1
MESSAGE_QUEUE_PORT=5672
```

使用 make 執行指令，啟動資料庫、RabbitMQ。

```
make create-mysql
make create-rabbitmq
```

安裝 Package

```
make install-python-env
```

發送任務

```
make sent-taiwan-stock-price-task
```

PROBLEMS OUTPUT DEBUG CONSOLE TERMINAL   1: bash

sam@DESKTOP-IKT69L5:~/FinMindBook/DataEngineering/Chapter5/5.5/5.5.5$ make sent-taiwan-stock-price-task
pipenv run python financialdata/producer.py taiwan_stock_price 2021-04-01 2021-04-12
Loading .env environment variables...
2021-10-30 17:07:52.336 | INFO | __main__:Update:20 - taiwan_stock_price, {'date': '2021-04-01', 'data_source': 'twse'}
2021-10-30 17:07:52.558 | INFO | __main__:Update:20 - taiwan_stock_price, {'date': '2021-04-01', 'data_source': 'tpex'}
2021-10-30 17:07:52.578 | INFO | __main__:Update:20 - taiwan_stock_price, {'date': '2021-04-02', 'data_source': 'twse'}
2021-10-30 17:07:52.580 | INFO | __main__:Update:20 - taiwan_stock_price, {'date': '2021-04-02', 'data_source': 'tpex'}
2021-10-30 17:07:52.581 | INFO | __main__:Update:20 - taiwan_stock_price, {'date': '2021-04-03', 'data_source': 'twse'}
2021-10-30 17:07:52.582 | INFO | __main__:Update:20 - taiwan_stock_price, {'date': '2021-04-03', 'data_source': 'tpex'}
2021-10-30 17:07:52.582 | INFO | __main__:Update:20 - taiwan_stock_price, {'date': '2021-04-06', 'data_source': 'twse'}
2021-10-30 17:07:52.583 | INFO | __main__:Update:20 - taiwan_stock_price, {'date': '2021-04-06', 'data_source': 'tpex'}
2021-10-30 17:07:52.584 | INFO | __main__:Update:20 - taiwan_stock_price, {'date': '2021-04-07', 'data_source': 'twse'}
2021-10-30 17:07:52.584 | INFO | __main__:Update:20 - taiwan_stock_price, {'date': '2021-04-07', 'data_source': 'tpex'}
2021-10-30 17:07:52.585 | INFO | __main__:Update:20 - taiwan_stock_price, {'date': '2021-04-08', 'data_source': 'twse'}
2021-10-30 17:07:52.586 | INFO | __main__:Update:20 - taiwan_stock_price, {'date': '2021-04-08', 'data_source': 'tpex'}
2021-10-30 17:07:52.586 | INFO | __main__:Update:20 - taiwan_stock_price, {'date': '2021-04-09', 'data_source': 'twse'}
2021-10-30 17:07:52.587 | INFO | __main__:Update:20 - taiwan_stock_price, {'date': '2021-04-09', 'data_source': 'tpex'}
2021-10-30 17:07:52.588 | INFO | __main__:Update:20 - taiwan_stock_price, {'date': '2021-04-10', 'data_source': 'twse'}
2021-10-30 17:07:52.588 | INFO | __main__:Update:20 - taiwan_stock_price, {'date': '2021-04-10', 'data_source': 'tpex'}

啟動 Celery

```
make run-celery-twse
```

這裡 Celery 多出一些額外參數：

concurrency
根據官方文件：

> *The number of worker processes/threads can be changed using the --concurrency argument and defaults to the number of CPUs available on the machine.*

意思是同時執行任務的線程數量預設為 CPU 核心數，由於證交所、櫃買中心會 ban IP，因此這裡需額外設定，同時只能做一個任務。

hostname
Worker 的名字，%h 代表的是 hostname。

-Q

代表這個 Celery 會去執行哪些任務列隊，例如 twse/tpex，分別代表這個 Celery 只會執行 twse/tpex 列隊的任務。

而為什麼這裡需要分兩個 Celery？因為實際上，一台機器可以同時執行證交所、櫃買中心爬蟲，因為這是兩個不同網站，同時爬蟲不用擔心太頻繁被 ban。

基本上只要啟動 Celery 後，再發送任務，就會順利將資料爬進 DB 囉。

5.6 定時爬蟲

5.6.1 為什麼需要定時爬蟲？

目前架構上，都是在 command line 上手動發送任務，這樣如果需要每天最新資料，不就需要每天手動執行 command line 了嗎？這非常沒效率，可能部分讀者會使用 crontab，但 crontab 不好做進版控裡，本書希望一切都使用版本控制，以下將介紹 Python 排程管理工具 APScheduler。

5.6.2 APScheduler

APScheduler 是一個 Python 的工作排程工具，由於是 Python 程式碼，可以做進版控，也不限於 Windows 或 Linux，你不用擔心 Windows 環境，沒有 crontab，以下提供一個簡單範例，也可從以下連結下載：

https://github.com/FinMind/FinMindBook/blob/master/DataEngineering/Chapter5/5.6/5.6.2/scheduler.py

```
scheduler.py
```

```python
import time

from apscheduler.schedulers.background import BackgroundScheduler
from loguru import logger


def sent_crawler_task(dataset: str):
    # 將此段，改成發送任務的程式碼
    logger.info(f"sent_crawler_task {dataset}")


def main():
    scheduler = BackgroundScheduler(timezone="Asia/Taipei")
    # 與crontab類似，設定何時執行，有小時、分鐘、秒參數，* 星號代表任意時間點
    scheduler.add_job(
        id="sent_crawler_task",
        func=sent_crawler_task,
        trigger="cron",
        hour="*",
        minute="*",
        day_of_week="*",
        second="*/5",
        args=["taiwan_stock_price"],
    )
    logger.info("sent_crawler_task")
    scheduler.start()


if __name__ == "__main__":
    main()
    while True:
        time.sleep(600)
```

基本上只要將以上 *logger.info(f"sent_crawler_task {dataset}")* 部分,改成
上一章節 update 的程式碼,就能每天定時爬蟲了。

執行方法

```
pipenv run python scheduler.py
```

接下來將 APScheduler 結合 5.5.5 的爬蟲程式。

5.6.3 APScheduler & 分散式爬蟲架構

APScheduler 結合爬蟲非常簡單,如下:

https://github.com/FinMind/FinMindBook/blob/master/DataEngineering/
Chapter5/5.6/5.6.3/financialdata/scheduler.py

```
scheduler.py
import time
import datetime

from apscheduler.schedulers.background import (
    BackgroundScheduler,
)
from financialdata.producer import (
    Update,
)
from loguru import logger


def sent_crawler_task():
    # 將此段,改成發送任務的程式碼
    # logger.info(f"sent_crawler_task {dataset}")
```

```
today = (
    datetime.datetime.today()
    .date()
    .strftime("%Y-%m-%d")
)
Update(
    dataset="taiwan_stock_price",
    start_date=today,
    end_date=today,
)


def main():
    scheduler = BackgroundScheduler(
        timezone="Asia/Taipei"
    )
    # 與crontab類似，設定何時執行，有小時、分鐘、秒參數，* 星號代表任意時間點
    scheduler.add_job(
        id="sent_crawler_task",
        func=sent_crawler_task,
        trigger="cron",
        hour="15",
        minute="0",
        day_of_week="mon-fri",
    )
    logger.info("sent_crawler_task")
    scheduler.start()


if __name__ == "__main__":
    main()
    while True:
        time.sleep(600)
```

這裡 Update 的 start_date、end_date 都設今天，就可以每天更新了，之後只需要手動做一次更新歷史資料即可，如下。

```
pipenv run python financialdata/producer.py taiwan_stock_price 2000-01-
01 2021-06-01
```

另外，scheduler 設定每周一到五，15 點執行，這時基本上證交所、櫃買中心資料都已經更新，如果不放心，可以設 15-17，這樣 15-17 每個整點都會發送一次爬蟲任務，因為 MySQL 有設定 PRIMARY KEY，因此也不用擔心資料重複 insert 問題。

以上是台股股價範例，你可以另外再加上期貨等等的其他爬蟲，基本上架構是一樣的。

本書到此，使用雲端，分散式進行爬蟲，且定時爬蟲。現在假設已經將所有歷史資料，都存入資料庫，那要如何取出呢？如果只有自己存取，寫幾段 SQL 可能就解決了，要給其他人使用呢？基本上，不可能讓對方直接對資料庫做連線，這有資安問題，API 是個不錯的選擇，只要架設好，就可以輕易的給其他人使用，且不限語言，以下將介紹 API 的設計。

06

資料提供─
RESTful API 設計

6.1 什麼是 API ？

API 是一種界接介面,用於不同程式、軟體之間的呼叫、傳遞,自由度高,你可以在 API 做一些限制,對方必須遵守你的規則,才能使用。

而 RESTful API 是一種設計規範,定義了 GET、POST、PUT、DELETE 等操作,例如拿取資料用 GET,推送資料用 POST,更新資料用 PUT,刪除資料用 DELETE,以下將一步步實作 RESTful API。

6.2 輕量 API --- Flask

6.2.1 建立第一個 API

以下是 Flask 官方介紹：

Flask is a lightweight WSGI web application framework

簡單來說，Flask 是一個輕量的 Web 框架，可以用於網站架設或是 API 架設，首先，使用 Flask 建立第一個 API。

安裝 package

```
pipenv install flask==1.0.0 PyJWT==1.7.1
```

https://github.com/FinMind/FinMindBook/blob/master/DataEngineering/
Chapter6/6.2/app.py

```python
app.py

from flask import Flask, jsonify


app = Flask(__name__)


@app.route("/", methods=["GET"])
def hello():
    return jsonify({"Hello": "World"})


if __name__ == "__main__":
    app.run(host="0.0.0.0", port=8888)
```

一樣先安裝 Package

```
pipenv sync
```

執行以下指令

```
pipenv run python app.py
```

會顯示以下畫面

```
PROBLEMS    OUTPUT    DEBUG CONSOLE    TERMINAL
sam@DESKTOP-IKT69L5:~/FinMindBook/DataEngineering/Chapter6/6.2$ pipenv run python app.py
 * Serving Flask app "app" (lazy loading)
 * Environment: production
   WARNING: Do not use the development server in a production environment.
   Use a production WSGI server instead.                                1
 * Debug mode: off
 * Running on http://0.0.0.0:8888/  (Press CTRL+C to quit)
                                          2
```

接著到 http://127.0.0.1:8888/，會看到 {"Hello": "World"}，代表成功架設了。

這時再開一個 Python，執行以下指令：

```
import requests

res = requests.get("http://127.0.0.1:8888/")
res.json()
```

就會出現 {"Hello": "World"}

```
PROBLEMS    OUTPUT    DEBUG CONSOLE    TERMINAL
sam@DESKTOP-IKT69L5:~/FinMindBook/DataEngineering/Chapter6/6.2$ python
Python 3.6.3 |Anaconda, Inc.| (default, Oct 13 2017, 12:02:49)
[GCC 7.2.0] on linux
Type "help", "copyright", "credits" or "license" for more information.
>>> import requests
>>>
... res = requests.get("http://127.0.0.1:8888/")          3
>>> res.json()
{'Hello': 'World'}                                          4
>>>
```

代表 API 架設成功，其他使用者，可以發 request 到以上架設的 API。

Flask 主打的是輕量 Web 框架，並非專門用於 API 架設，那麼接下來，介紹專門用於高效能的 API，Fastapi，它擁有完整的 API 架構，包含良好的架構設計、WebSocket、測試等。

6.3 高效能 API --- FastAPI

6.3.1 建立第一個 API

同樣，引用 FastAPI 官方介紹：

FastAPI is a modern, fast (high-performance), web framework for building APIs with Python 3.6+ based on standard Python type hints.

FastAPI 是一個高效能的 API 框架，快速、簡單，先來一個簡單的例子。安裝 Pacakge

```
pipenv install fastapi==0.63.0 uvicorn==0.14.0
```

程式碼可以從以下連結下載：

https://github.com/FinMind/FinMindBook/blob/master/DataEngineering/
Chapter6/6.3/6.3.1/main.py

```
main.py
from fastapi import FastAPI


app = FastAPI()



@app.get("/")
def read_root():
    return {"Hello": "World"}



@app.get("/taiwan_stock_price")
def taiwan_stock_price():
    print("get data from mysql")
    return {"data": 123}
```

執行以下指令

```
pipenv run uvicorn main:app --reload --port 8888
```

成功的話，會顯示以下畫面：

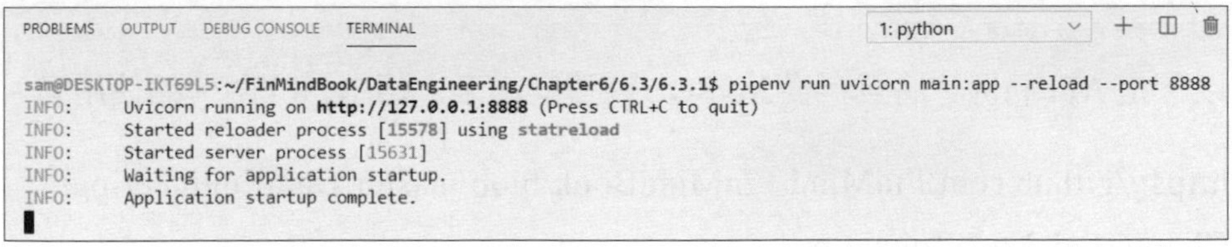

接著到 http://127.0.0.1:8888/，就會顯示 {"Hello":"World"}。這時模擬使用者，要對這個 API 發送 request，如果畫面與下圖相同，代表成功模擬了，之後只要把 get data from mysql 的部分，改成實際的程式碼，就能完整的串接 API，拿到資料了，下個章節將做個示範。

```
PROBLEMS    OUTPUT    DEBUG CONSOLE    TERMINAL

sam@DESKTOP-IKT69L5:~/FinMindBook/DataEngineering/Chapter6/6.3/6.3.2$ python
Python 3.6.3 |Anaconda, Inc.| (default, Oct 13 2017, 12:02:49)
[GCC 7.2.0] on linux
Type "help", "copyright", "credits" or "license" for more information.
>>> import requests
>>> res = requests.get("http://127.0.0.1:8888/taiwan_stock_price")
>>> res.json()
{'data': 123}
>>> ▮
```

6.3.2　建立 API 回傳真實資料

首先，使用 docker ps 檢查，確保前面爬蟲用到的 MySQL Docker 活著。

```
sam@DESKTOP-IKT69L5:~/FinMindBook/DataEngineering/Chapter6/6.3/6.3.2$ docker ps
CONTAINER ID    IMAGE                            COMMAND                 CREATED        STATUS          PORTS
                                                                                                        NAMES
6ab83448cc06    phpmyadmin/phpmyadmin:5.1.0      "/docker-entrypoint...."  6 hours ago    Up 6 hours      0.0.0.0:8000->80/tcp, :::8000-
>80/tcp                                                                                                  53_phpmyadmin_1
bc9a660d7186    mysql:8.0                        "docker-entrypoint.s…"   6 hours ago    Up 6 hours      0.0.0.0:3306->3306/tcp, :::330
6->3306/tcp, 33060/tcp                                                                                  53_mysql_1
2379b4802294    mher/flower:0.9.5                "flower flower --bro…"   8 hours ago    Up 1 second     0.0.0.0:5555->5555/tcp, :::555
5->5555/tcp                                                                                             553_flower_1
27ef49d1f49e    rabbitmq:3.6-management-alpine   "docker-entrypoint.s…"   8 hours ago    Up 1 second     4369/tcp, 5671/tcp, 0.0.0.0:56
72->5672/tcp, :::5672->5672/tcp, 15671/tcp, 25672/tcp, 0.0.0.0:15672->15672/tcp, :::15672->15672/tcp    553_rabbitmq_1
```

接著安裝 API 所需套件。

```
pipenv install fastapi==0.63.0 pandas==1.1.5 pymysql==1.0.2
sqlalchemy==1.4.20 uvicorn==0.14.0
```

以下是 API 範例，將資料從 MySQL 撈資料出來，藉由 API 回傳給使用者。

https://github.com/FinMind/FinMindBook/blob/master/DataEngineering/
Chapter6/6.3/6.3.2/main.py

main.py
```
import pandas as pd
from fastapi import FastAPI
from sqlalchemy import create_engine, engine
```

```python
def get_mysql_financialdata_conn() -> engine.base.Connection:
    address = "mysql+pymysql://root:test@127.0.0.1:3306/financialdata"
    engine = create_engine(address)
    connect = engine.connect()
    return connect


app = FastAPI()


@app.get("/")
def read_root():
    return {"Hello": "World"}


@app.get("/taiwan_stock_price")
def taiwan_stock_price(
    stock_id: str = "",
    start_date: str = "",
    end_date: str = "",
):
    sql = f"""
    select * from TaiwanStockPrice
    where StockID = '{stock_id}'
    and Date>= '{start_date}'
    and Date<= '{end_date}'
    """
    mysql_conn = get_mysql_financialdata_conn()
    data_df = pd.read_sql(sql, con=mysql_conn)
    data_dict = data_df.to_dict("records")
    return {"data": data_dict}
```

以上 API 的設計，讓使用者可以輸入股票代碼 stock_id、開始日期 start_date、結束日期 end_date，之後 API 接收到這些參數後，再去資料庫撈資料，回傳給使用者。

同樣，模擬使用者，對 API 發 request，結果如下，使用者成功拿到 data

啟動 API Server。

```
pipenv run uvicorn main:app --reload --port 8888
```

模擬 user

```
sam@DESKTOP-IKT69L5:~/FinMindBook/DataEngineering/Chapter6/6.3/6.3.2$ python
Python 3.6.3 |Anaconda, Inc.| (default, Oct 13 2017, 12:02:49)
[GCC 7.2.0] on linux
Type "help", "copyright", "credits" or "license" for more information.
>>> import requests
>>> import pandas as pd
>>>
>>> payload = dict(
...     stock_id="2330",
...     start_date="2021-04-01",
...     end_date="2021-04-15",
... )
>>> res = requests.get("http://127.0.0.1:8888/taiwan_stock_price", params=payload)
>>> df = pd.DataFrame(res.json()["data"])
>>> df
  StockID  TradeVolume  Transaction    TradeValue   Open    Max    Min  Close  Change        Date
0    2330     45972766        48170  27520742963  598.0  602.0  594.0  602.0    15.0  2021-04-01
1    2330     37664216        42422  23045132094  615.0  616.0  608.0  610.0     8.0  2021-04-06
2    2330     28140964        28395  17175949080  614.0  614.0  608.0  610.0     0.0  2021-04-07
3    2330     26658283        32098  16217570357  606.0  613.0  603.0  613.0     3.0  2021-04-08
4    2330     31601619        28920  19344167081  618.0  618.0  609.0  610.0    -3.0  2021-04-09
>>>
```

到這步驟，API 基本上就完成第一階段了，是不是很簡單阿，接下來只要把 api 部屬在雲端機器上，使用者就能根據雲端 IP，對你的 API 發送 request，拿到資料。

那實際上 FastAPI 效能，到底有沒有比 Flask 好呢？以下章節將使用 ApacheBench 這個壓力測試的工具，對 API 做測試。

6.3.3 ApacheBench 壓力測試

根據官方文件介紹：

> *ab is a tool for benchmarking your Apache Hypertext Transfer Protocol (HTTP) server. It is designed to give you an impression of how your current Apache installation performs. This especially shows you how many requests per second your Apache installation is capable of serving.*

簡單來說呢，ApacheBench（簡稱 ab），是一個對伺服器進行基準測試的工具，可以模擬多個使用者，同時對伺服器發送多個 request，最後統計每次 request 需要多少時間，是否中間有 failed，以下將做個範例，分別對 Flask、FastAPI 做壓測，而 Flask、FastAPI 皆使用官方 GitHub 推薦的方法做啟動。

安裝

```
apt-get install apache2-utils -y
```

壓測指令如下：

```
ab -c 10 -n 1000  'http://127.0.0.1:8888/'
```

首先對 Flask 做壓測，結果如下，只看幾個重要指標就好。

```
Complete requests:       1000
Failed requests:         0
Total transferred:       164000 bytes
HTML transferred:        18000 bytes
Requests per second:     476.31 [#/sec] (mean)
Time per request:        20.995 [ms] (mean)
Time per request:        2.099 [ms] (mean, across all concurrent requests)
Transfer rate:          76.28 [Kbytes/sec] received
```

接著對 FastAPI 做壓測，結果如下：

```
Complete requests:       1000
Failed requests:         0
Total transferred:       142000 bytes
HTML transferred:        17000 bytes
Requests per second:     789.24 [#/sec] (mean)
Time per request:        12.670 [ms] (mean)
Time per request:        1.267 [ms] (mean, across all concurrent requests)
Transfer rate:          109.44 [Kbytes/sec] received
```

首先，介紹幾個統計指標：

Complete requests 代表完成幾個 request，這裡代表 1000 個都完成。

Failed requests 代表有幾個 request 失敗，如果你的伺服器不穩定，在高壓情況下，可能發生 1% 的失敗，這時 ApacheBench 就能很好的模擬出情境。

Total transferred 代表總共傳輸量，這裡 FastAPI 比 Flask 好約 20%。

Requests per second 代表每秒能承受多少 request。

Time per request 這裡有兩個，以 FastAPI 為例，分別是 12.670 ms、1.267 ms，差了 10 倍，剛好對應到，模擬 10 個使用者，第一個 12.670 ms，是這 10 個 user 的平均回應時間，第二個 1.267 ms，是每一個 user 的平均回應時間。FastAPI 跟 Flask 比較，快了 40%。

以上只是初步比較，當然 FastAPI 還有架構設計上的優點，這邊先不展示，且 FastAPI 有額外的功能，自動產生 Swagger 文件，這非常好用，下個章節將展示此功能。

6.3.4 API 介面 --- Swagger

Swagger 是一種自動生成的 OpenAPI 文件，可以直接在 Web 上對 API 發送 request，就不用寫 code 了，這對於開發上，有很大的幫助，先來看前面 FastAPI 自動生成的文件，到 http://127.0.0.1:8888/docs 以上網址，會顯示以下畫面：

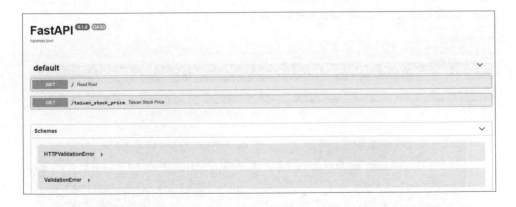

這就是前面使用 FastAPI 架設所有的 API，因為前面章節，只設定兩種 API，因此只顯示兩種，接著來試試看用在網頁上，對 API 發送 request。

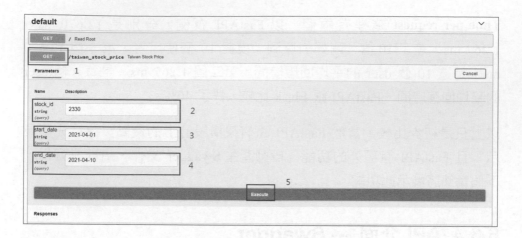

第一，先點 taiwan_stock_price 的 GET，之後會展開出輸入頁面，包含 stock_id、start_date、end_date 這三個參數，這也就是 API 程式中，定義需要輸入的參數，本書使用的框架，直接將程式碼轉換在 Web 介面。輸入完之後，再點 Execute 按鈕，就會發送 request 了，結果如下：

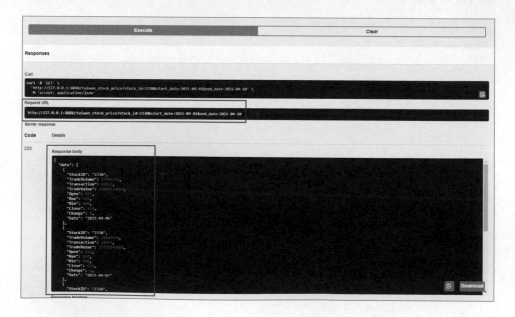

Request URL 的部分，就是網頁發送的 URL，包含帶的參數，而 Response body，就是 API 回應的結果，可以看到有股價相關的資料，這時相對於使用程式，有網站就能夠輕易的操作 API，非常方便，這也是 FastAPI 相較於 Flask 的優點之一。

本書到此，已經能架設 API 了，剩下的只需要將程式在雲端 Linode 上執行，修改一些設定，把 localhost、127.0.0.1 改成實體 IP，就能架設一個 API 讓別人來使用你的服務了，甚至還有 Swagger 介面，讓使用者清楚知道你的 API 有哪些功能。

目前爬蟲、API 已經完成第一步了，剩下只要持續新增爬蟲、API，就能讓資料越來越豐富，下個章節，將使用 Docker 包裝爬蟲、API。

Docker 主要用於後續持續開發的爬蟲、API，方便持續新版本更新迭代，試想一個情境，當未來 API 使用超過 2 台機器做分散，爬蟲也使用 2 台機器做分散，這時就有 4 台機器了，那更新一次爬蟲、API，需要連線到 4 台機器上，非常麻煩。而照著本書的方法，後續只需要按一個按鈕，就可以自動更新，當然自動化前期，需要花一點時間做開發，以下將一一介紹整個過程。

容器管理工具 Docker

7.1 為什麼要用 Docker

前面章節的 MySQL、RabbitMQ 都用 Docker，那為什麼爬蟲跟 API 也需要用 Docker 呢？因為實務開發上，可能遇到以下幾個情境。

1. **開發環境與生產環境不完全相同**。在產品開發上，這是很常見的情境，團隊多個成員，可能同時用 Mac、Windows、Linux 做開發，雲端基本上都是 Linux 環境，因此開發的程式碼，在本地端可以順利運行，不保證在雲端上能正常，即使開發上也是 Linux，Linux 也有多種版本，而 Docker 是獨立環境，不論在 Windows、Linux、Mac，都不受影響，因此 Docker 可以確保開發與生產的穩定性。

2. **版本更動**。在開發過程中，會持續新增爬蟲、API，因此生產環境的版本，也會不斷更動，特別使用分散式架構，如果有 10 台機器，那換版過程將非常複雜，使用 Docker，再加上 Chapter 9 的 CI/CD，可以做到一鍵換版。

3. **拓展性**。本書使用分散式架構爬蟲，目前約 3-5 台機器，未來可能 30-50 台，因此需考量拓展性，再加上 API、資料庫未來也會分散式，用 Loading Balance 做分流，因此需要可高拓展的架構，Docker 就是高拓展性架構的基底。

4. **多服務、大型架構**。以數據分析來說，除了 API、爬蟲，將分析結果視覺化，是很常見的，因此需要 Web 才能展現給多數人，雖然 Python 有 Web 的視覺化工具可以使用（如以 plotly 為基底的 dash），但高可用的 Web 基本上會走前後端分離，因此持續開發，除了爬蟲、API 外，還會需要許多服務。以 FinMind 為例，有爬蟲、API、分析、Backend、Web、DataBase、Airflow、Log 等多個服務，因此都用 Docker，可以很容易地跟其他服務做對接。

5. **K8s**。以現代架構來說，K8s 算是必備，那要如何學習 K8s？可以先從 Docker 開始，讓目前的服務容器化、微服務化，未來架構走到 K8s，相對來說容易。

Docker 大致上有以上幾點好處，主要都是考量未來的架構設計，這也是筆者真實的經驗，真實體會過 Docker 的好處。下圖是 FinMind 目前的產品架構，一開始也是從爬蟲、API 開始，慢慢衍生出許多服務，而基本上整個產品都圍繞在 Docker 上，這也是為什麼本書一開始就先介紹 Docker，Docker 幫筆者省去非常多時間。

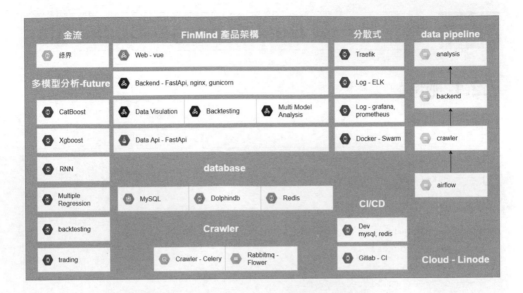

以初學者來說，一開始開發產品時，可能不會接觸到這些，這需要花費不少時間踩雷，累積經驗，即使已經工作一段時間，由於公司分工細，你很難接觸到全貌，無法學習到某些技能，這是優點也是缺點，即使有接觸架構，可能也是接續前人的基礎，不過最難的還是一個真實場景，高流量會遇到完全不同的問題，也會考慮高效的設計架構，而本書直接提供以 FinMind 產品為例的架構，可以做初步參考，讓讀者在工作之餘，有額外管道學習這些技能。

FinMind 專注在多種金融數據上，同樣的方式，讀者可以嘗試在不同領域的數據上實現。以下章節將著重在架構層面上，關於爬蟲、API 的程式碼，在上面的章節已做過介紹，剩下只需花時間持續開發。

7.2 建立第一個 Docker Image--Dockerfile

7.2.1 爬蟲

架構如下：

```
├── Dockerfile
├── Makefile
├── Pipfile
├── Pipfile.lock
├── README.md
├── cache.Dockerfile
├── crawler.yml
├── crawler_multi_celery.yml
├── create_partition_table.sql
├── docker-compose.yml
├── financialdata
│   ├── __init__.py
│   ├── backend
│   │   ├── __init__.py
│   │   └── db
│   │       ├── __init__.py
│   │       ├── clients.py
│   │       ├── db.py
│   │       └── router.py
│   ├── config.py
│   ├── crawler
│   │   ├── __init__.py
│   │   ├── taiwan_futures_daily.py
│   │   └── taiwan_stock_price.py
│   ├── producer.py
│   ├── scheduler.py
│   ├── schema
│   │   ├── __init__.py
│   │   └── dataset.py
│   └── tasks
```

```
│          ├── __init__.py
│          ├── task.py
│          └── worker.py
├── genenv.py
├── local.ini
├── mysql.yml
├── rabbitmq.yml
├── scheduler.yml
└── setup.py
```

要建立 Docker Image，基本上都是使用 Dockerfile 這個文件，首先，將爬蟲的程式碼，包成 Docker Image，使用以下 Dockerfile：

https://github.com/FinMind/FinMindBook/blob/master/DataEngineering/
Chapter7/7.2/7.2.1/Dockerfile

```
Dockerfile
# 由於 continuumio/miniconda3:4.3.27 中的 Debian
# 版本太舊，因此改用 ubuntu 系統
FROM ubuntu:18.04

# 系統升級、安裝 python
RUN apt-get update && apt-get install python3.6 -y && apt-get install
python3-pip -y

RUN mkdir /FinMindProject
COPY . /FinMindProject/
WORKDIR /FinMindProject/

# env
ENV LC_ALL=C.UTF-8
ENV LANG=C.UTF-8
```

```
# install package
RUN pip3 install pipenv==2020.6.2
RUN pipenv sync


# genenv
RUN VERSION=RELEASE python3 genenv.py
```

先介紹幾個常用指令。

From。以 Image 為基底,在這上面去建立你的 Docker 環境。

Run。執行設定的指令,你可以使用此指令,去安裝所需的工具。

COPY。將本地端的檔案,複製進 Docker Image 裡面。

ENV。設定環境變數。

WORKDIR。設定你的工作目錄。

接著來一步步解析 Dockerfile 為什麼這樣寫:

1. 使用 ubuntu:18.04 為基底,從乾淨的環境開始建立環境,未來如果 ubuntu 18.04 不支援,要升級到 20.04、22.04 也比較方便。

```
FROM ubuntu:18.04
```

2. 系統升級、安裝 python。

```
RUN apt-get update && apt-get install python3.6 -y && apt-get install
python3-pip -y
```

3. 建立資料夾,待會存放爬蟲的程式碼。

```
RUN mkdir /FinMindProject
```

4. 將當前目錄下的檔案，複製進去，這裡使用點 . ，代表當前路徑所有
檔案，複製進去 FinMindProject 這個資料夾。

```
COPY . /FinMindProject/
```

5. 設定工作目錄在 FinMindProject 底下，因為需要啟動 Celery 做爬蟲，
因此設定在這目錄底下。

```
WORKDIR /FinMindProject/
```

6. 設定環境變數。

```
ENV LC_ALL=C.UTF-8
ENV LANG=C.UTF-8
```

7. 安裝 pipenv 且安裝所需 Package。

```
RUN pip3 install pipenv==2020.6.2
RUN pipenv sync
```

8. 建立環境變數。

```
RUN VERSION=RELEASE python3 genenv.py
```

以上指令就能能一步步安裝出所需的 Docker 環境，而初學遇到最大的問
題是，不知如何寫起，這時可以先用以下指令 run docker：

```
docker run -it --rm ubuntu:18.04 bash
```

成功後會進入到 Docker Container 容器內。

```
sam@DESKTOP-IKT69L5:~/FinMindBook$ docker run -it --rm ubuntu:18.04 bash
root@147ee41ad5eb:/# ▮
```

再一步步嘗試指令，並記錄在 Dockerfile，最後找出 SOP，根據 SOP 流程，就能建立起所需環境。

而 Dockerfile 建立好了之後，使用以下指令，建立 Docker Image：

```
docker build -f Dockerfile -t crawler:7.2.1 .
```

-f Dockerfile，是指定使用 Dockerfile 這個檔案，去建立 Image，如果你使用其他檔名，就在這更動檔名。

-t crawler:7.2.1，是指建立出的 Image 名稱。

.，就是當前目錄。

建立完 Docker Image 之後，使用 **Docker Images** 指令，就會多一個你剛剛建立的 Image：

```
sam@DESKTOP-IKT69L5:~/FinMindBook/DataEngineering/Chapter7/7.2/7.2.1$ docker images
REPOSITORY              TAG                     IMAGE ID        CREATED         SIZE
crawler                 7.2.1                   c5abc9f89a5b    16 seconds ago  866MB
```

這時有了爬蟲的 Docker Image 後，接著來用 Docker 啟動 Celery，而啟動方法與啟動 MySQL、RabbitMQ 的方法相同，Docker-Compose，以下是爬蟲的 Docker-Compose：

https://github.com/FinMind/FinMindBook/blob/master/DataEngineering/ Chapter7/7.2/7.2.1/docker-compose.yml

```
docker-compose.yml

version: '3.0'
services:
  crawler_twse:
    image: crawler:7.2.1
    hostname: "twse"
```

```
    command: pipenv run celery -A financialdata.tasks.worker worker
--loglevel=info --concurrency=1  --hostname=%h -Q twse
    restart: always
    environment:
      - TZ=Asia/Taipei
    networks:
        - dev

networks:
  dev:
```

使用以下指令：

```
docker-compose -f docker-compose.yml up
```

這時會出現以下錯誤：

> *(2003, "Can't connect to MySQL server on '127.0.0.1' ([Errno 111] Connection refused)")*

你應該會很疑惑，你的電腦中，明明有啟動 MySQL，也可以透過 127.0.0.1 去做連線，為什麼會顯示無法連線？這是因為，Docker 內部是獨立環境，連 Network 網路也走獨立網路，那這時要如何讓 Docker 網路相通呢？這時需要先建立 Docker Network，指令如下：

```
docker network create my_network
```

建立了 my_network Docker 網路，接著要額外設定，MySQL、RabbitMQ 都使用這個網路，需要修改 Docker-Compose 設定。

https://github.com/FinMind/FinMindBook/blob/master/DataEngineering/ Chapter7/7.2/7.2.1/mysql.yml

```
mysql.yml
version: '3'
services:

  mysql:
      image: mysql:8.0
      command: mysqld --default-authentication-plugin=mysql_native_password
      ports:
          - 3306:3306
      environment:
          MYSQL_DATABASE: mydb
          MYSQL_USER: user
          MYSQL_PASSWORD: test
          MYSQL_ROOT_PASSWORD: test
      volumes:
          - mysql:/var/lib/mysql
      networks:
          - my_network

  phpmyadmin:
      image: phpmyadmin/phpmyadmin:5.1.0
      links:
          - mysql:db
      ports:
          - 8000:80
      depends_on:
        - mysql
      networks:
          - my_network

networks:
  my_network:
```

```
    # 加入已經存在的網路
    external: true


volumes:
   mysql:
      external: true
```

這邊 Network 改為 my_network，並使用 external: true，讓 Docker-Compose 在啟動時，選擇加入現成的網路 my_network，RabbitMQ 也是同樣動作。

https://github.com/FinMind/FinMindBook/blob/master/DataEngineering/
Chapter7/7.2/7.2.1/rabbitmq.yml

```
rabbitmq.yml
version: '3'
services:

  rabbitmq:
    image: 'rabbitmq:3.6-management-alpine'
    ports:
      - '5672:5672'
      - '15672:15672'
    environment:
      RABBITMQ_DEFAULT_USER: "worker"
      RABBITMQ_DEFAULT_PASS: "worker"
      RABBITMQ_DEFAULT_VHOST: "/"
    networks:
      - my_network

  flower:
    image: mher/flower:0.9.5
    command: ["flower", "--broker=amqp://worker:worker@rabbitmq",
```

```
"--port=5555"]
    ports:
      - 5555:5555
    depends_on:
      - rabbitmq
    networks:
      - my_network

networks:
  my_network:
    # 加入已經存在的網路
    external: true
```

這時先關閉之前啟動的 MySQL、RabbitMQ。

先使用

```
docker ps
```

查看目前啟動的 Docker：

```
sam@DESKTOP-IKT69L5:~/FinMindBook/DataEnginner/Chapter5/5.5/5.5.5$ docker ps
CONTAINER ID   IMAGE                              COMMAND                  CREATED       STATUS            PORTS
                                                                                         NAMES
c70c63dd2b06   phpmyadmin/phpmyadmin:5.1.0        "/docker-entrypoint..."  8 seconds ago    Up 6 seconds     0.0.0.0:8000->80/tcp, :::8000->80/tcp
                                                                                         555_phpmyadmin_1
06a555086c7b   mysql:8.0                          "docker-entrypoint.s…"   9 seconds ago    Up 7 seconds     0.0.0.0:3306->3306/tcp, :::3306->3306/tcp, 33060/
tcp                                                                                      555_mysql_1
0ad16ebb2ca9   mher/flower:0.9.5                  "flower flower --bro…"   26 seconds ago   Up 23 seconds    0.0.0.0:5555->5555/tcp, :::5555->5555/tcp
                                                                                         555_flower_1
f5f422f299e8   rabbitmq:3.6-management-alpine     "docker-entrypoint.s…"   28 seconds ago   Up 26 seconds    4369/tcp, 5671/tcp, 0.0.0.0:5672->5672/tcp, :::56
72->5672/tcp, 15671/tcp, 25672/tcp, 0.0.0.0:15672->15672/tcp, :::15672->15672/tcp   555_rabbitmq_1
```

使用 docker stop 關閉 Container：

```
docker stop 555_flower_1 555_mysql_1 555_phpmyadmin_1 555_rabbitmq_1
```

使用 docker rm 刪除 Container：

```
docker rm 555_flower_1 555_mysql_1 555_phpmyadmin_1 555_rabbitmq_1
```

接著啟動新的 MySQL、RabbitMQ：

```
docker-compose -f rabbitmq.yml up -d
docker-compose -f mysql.yml up -d
```

這時爬蟲需要修改 local.init 設定，才能成功在 Docker Network 成功連上
MySQL、RabbitMQ，如下（單純擷取 RELEASE 部分）。

local.ini

```
[RELEASE]
MYSQL_DATA_HOST = mysql
MYSQL_DATA_USER = root
MYSQL_DATA_PASSWORD = test
MYSQL_DATA_PORT = 3306
MYSQL_DATA_DATABASE = financialdata
WORKER_ACCOUNT = worker
WORKER_PASSWORD = worker
MESSAGE_QUEUE_HOST = rabbitmq
MESSAGE_QUEUE_PORT = 5672
```

特別注意 HOST 部分，在 Docker 內部，不是用 IP 做連線，而是使
用 service name，以目前 mysql、rabbitmq 來說，service name 分別是
mysql、rabbitmq，如下圖：

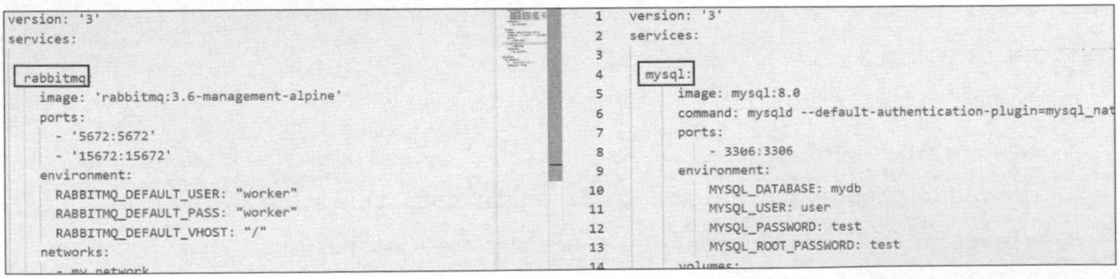

```
version: '3'                                    1  version: '3'
services:                                       2  services:
                                                3
  rabbitmq                                      4    mysql:
    image: 'rabbitmq:3.6-management-alpine'     5      image: mysql:8.0
    ports:                                      6      command: mysqld --default-authentication-plugin=mysql_nat
      - '5672:5672'                             7      ports:
      - '15672:15672'                           8        - 3306:3306
    environment:                                9      environment:
      RABBITMQ_DEFAULT_USER: "worker"          10        MYSQL_DATABASE: mydb
      RABBITMQ_DEFAULT_PASS: "worker"          11        MYSQL_USER: user
      RABBITMQ_DEFAULT_VHOST: "/"              12        MYSQL_PASSWORD: test
    networks:                                  13        MYSQL_ROOT_PASSWORD: test
      - my_network                            14      volumes:
```

用 service name 有幾點好處：

1. 你的 Docker 架構，可以很輕易的部屬在不同環境上，包含測試與產品環境上，不受 IP 限制。

2. 如果機器臨時需要維護，某些服務需要切換機器，如果用 IP，你所有服務都要改 IP，如果用 service name，那就不用改，方便很多。

3. 安全性，在多人開發中，如果 local.ini 包含實體 IP，你無法保證機器資訊會不會洩漏，因此使用 service name，即使對外洩漏，對方也無法知道，你的服務在哪台機器，只會知道，團隊使用的 MySQL host 是 mysql。

4. 拓展性，Loading Balance 分流，不過初期可能感受不到，也需要做額外設定，這裡先不提起，避免主題混淆。

修改完設定後，重新建立（build）Docker Image，並將 Docker-Compose 中的 networks 設定成 my_network：

https://github.com/FinMind/FinMindBook/blob/master/DataEngineering/Chapter7/7.2/7.2.1/crawler.yml

```
crawler.yml
version: '3.0'
services:
  crawler_twse:
    image: crawler:7.2.1
    hostname: "twse"
    command: pipenv run celery -A financialdata.tasks.worker worker
--loglevel=info --concurrency=1  --hostname=%h -Q twse
    restart: always
    environment:
```

```
      - TZ=Asia/Taipei
    networks:
        - my_network

networks:
  my_network:
    # 加入已經存在的網路
    external: true
```

接著啟動 Celery

```
docker-compose -f crawler.yml up
```

如果結果如下圖，代表成功了：

```
sam@DESKTOP-IKT69L5:~/FinMindBook/DataEngineering/Chapter7/7.2/7.2.1$ docker-compose -f crawler.yml up
WARNING: Found orphan containers (721_mysql_1, 721_phpmyadmin_1, 721_rabbitmq_1, 721_flower_1) for this project. If you removed or
renamed this service in your compose file, you can run this command with the --remove-orphans flag to clean it up.
Starting 721_crawler_twse_1 ... done
Attaching to 721_crawler_twse_1
crawler_twse_1  | Loading .env environment variables...
crawler_twse_1  |
crawler_twse_1  |  -------------- celery@twse v5.1.0 (sun-harmonics)
crawler_twse_1  | --- ***** -----
crawler_twse_1  | -- ******* ---- Linux-5.4.72-microsoft-standard-WSL2-x86_64-with-debian-8.5 2021-10-31 00:44:31
crawler_twse_1  | - *** --- * ---
crawler_twse_1  | - ** ---------- [config]
crawler_twse_1  | - ** ---------- .> app:         task:0x7f3733aeef60
crawler_twse_1  | - ** ---------- .> transport:   amqp://worker:**@rabbitmq:5672//
crawler_twse_1  | - ** ---------- .> results:     disabled://
crawler_twse_1  | - *** --- * ---- .> concurrency: 1 (prefork)
crawler_twse_1  | -- ******* ---- .> task events: OFF (enable -E to monitor tasks in this worker)
crawler_twse_1  | --- ***** -----
crawler_twse_1  |  -------------- [queues]
crawler_twse_1  |                .> twse             exchange=twse(direct) key=twse
crawler_twse_1  |
crawler_twse_1  |
crawler_twse_1  | [tasks]
crawler_twse_1  |   . financialdata.tasks.task.crawler
```

可以看到，**Connected to amqp://worker:**@rabbitmq:5672//**，RabbitMQ 連線成功，並且連線的 host 是 RabbitMQ，也就是 service name，代表成功加入到 my_network，是在同一個 Docker 網路中。

這時跳轉到 http://localhost:5555/

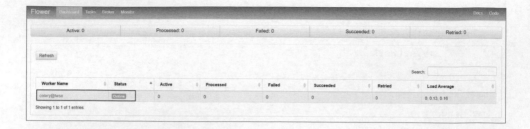

可以看到 Celery 成功連線上，不過這只有 twse，接著再把其他 Celery 加上去：

https://github.com/FinMind/FinMindBook/blob/master/DataEngineering/
Chapter7/7.2/7.2.1/crawler_multi_celery.yml

crawler_multi_celery.yml

```
version: '3.0'
services:
  crawler_twse:
    image: crawler:7.2.1
    hostname: "twse"
    command: pipenv run celery -A financialdata.tasks.worker worker
--loglevel=info --concurrency=1  --hostname=%h -Q twse
    restart: always
    environment:
      - TZ=Asia/Taipei
    networks:
      - my_network

  crawler_tpex:
    image: crawler:7.2.1
    hostname: "tpex"
    command: pipenv run celery -A financialdata.tasks.worker worker
--loglevel=info --concurrency=1  --hostname=%h -Q tpex
```

```
    restart: always
    environment:
      - TZ=Asia/Taipei
    networks:
        - my_network

  crawler_taifex:
    image: crawler:7.2.1
    hostname: "taifex"
    command: pipenv run celery -A financialdata.tasks.worker worker
--loglevel=info --concurrency=1  --hostname=%h -Q taifex
    restart: always
    environment:
      - TZ=Asia/Taipei
    networks:
        - my_network

networks:
  my_network:
    # 加入已經存在的網路
    external: true
```

接著啟動 Celery

```
docker-compose -f crawler_multi_celery.yml up
```

切換到 http://localhost:5555/，就可以看到，有三個 Celery。

Active: 0		Processed: 0		Failed: 0		Succeeded: 0		Retried: 0	

Refresh

Search:

Worker Name	Status	Active	Processed	Failed	Succeeded	Retried	Load Average
celery@tenx	Online	0	0	0	0	0	0.91, 0.27, 0.09
celery@taifex	Online	0	0	0	0	0	0.91, 0.27, 0.09
celery@tpex	Online	0	0	0	0	0	0.91, 0.27, 0.09

Showing 1 to 3 of 3 entries

這時成功將爬蟲程式，包成 Docker Image，並成功連上 RabbitMQ、MySQL。

下一步，將 APScheduler 排程，也用 Docker 部屬，yml 檔如下：

https://github.com/FinMind/FinMindBook/blob/master/DataEngineering/Chapter7/7.2/7.2.1/scheduler.yml

```
scheduler.yml
version: '3.0'
services:
  scheduler:
    image: crawler:7.2.1
    hostname: "twse"
    command: pipenv run python financialdata/scheduler.py
    restart: always
    environment:
      - TZ=Asia/Taipei
    networks:
        - my_network

networks:
  my_network:
    # 加入已經存在的網路
    external: true
```

使用以下指令啟動 scheduler docker container：

```
docker-compose -f scheduler.yml up
```

成功執行的畫面如下，這時時間到，就會自動發送 task 到 RabbitMQ 上，Worker 接收到任務，就會去執行爬蟲了。

```
sam@DESKTOP-IKT69L5:~/FinMindBook/DataEngineering/Chapter7/7.2/7.2.1$ docker-compose -f scheduler.yml up
WARNING: Found orphan containers (721_phpmyadmin_1, 721_crawler_twse_1, 721_rabbitmq_1, 721_mysql_1, 721_flower_1) for this project
. If you removed or renamed this service in your compose file, you can run this command with the --remove-orphans flag to clean it
up.
Creating 721_scheduler_1 ... done
Attaching to 721_scheduler_1
scheduler_1  | Loading .env environment variables...
scheduler_1  | 2021-10-31 00:45:46.504 | INFO     | __main__:main:27 - sent_crawler_task
```

當然如果你想讓他在背景執行，就執行以下指令：

```
docker-compose -f scheduler.yml up -d
```

不用擔心壞掉怎麼辦，在 scheduler.yml 中有設定：

```
restart: always
```

壞了會自動重啟。

下一章節，將 API 也包成 Docker Image。

7.2.2 API

架構如下：

```
├──── api
│    ├──── config.py
│    ├──── __init__.py
│    └──── main.py
├──── api.yml
├──── Dockerfile
├──── genenv.py
├──── local.ini
├──── Makefile
├──── Pipfile
├──── Pipfile.lock
├──── README.md
└──── setup.py
```

基本上與爬蟲是類似的，使用以下 Dockerfile 建立 Image：

https://github.com/FinMind/FinMindBook/blob/master/DataEngineering/
Chapter7/7.2/7.2.2/Dockerfile

```
Dockerfile

# 由於 continuumio/miniconda3:4.3.27 中的 Debian
# 版本太舊，因此改用 ubuntu 系統
FROM ubuntu:18.04

# 系統升級、安裝 python
RUN apt-get update && apt-get install python3.6 -y && apt-get install
python3-pip -y

RUN mkdir /FinMindProject
COPY . /FinMindProject/
WORKDIR /FinMindProject/

# env
ENV LC_ALL=C.UTF-8
ENV LANG=C.UTF-8

# install package
RUN pip3 install pipenv==2020.6.2
RUN pipenv sync

# genenv
RUN VERSION=RELEASE python3 genenv.py

# 預設執行的指令
CMD ["pipenv", "run", "uvicorn", "api.main:app", "--host", "0.0.0.0",
"--port", "8888"]
```

Dockerfile 與爬蟲的 SOP 是一樣的，先安裝好 Python 所需套件、環境變數後，比較特別的是，使用 CMD 設定預設執行的指令，這樣當 Run Docker Image 時，就會直接啟動 API，但為什麼在爬蟲時，不這樣寫呢？因為爬蟲使用 Celery，需要多種不同的 command，如 crawler_twse、crawler_tpex，因此 CMD 需要在 Docker-Compose 上設定，而 API 只有一種，因此寫在 Dockerfile 裡面，你也可以嘗試模仿爬蟲，將 CMD 寫在 Docker-Compose 中。

--host 0.0.0.0，是指定 API 不再用 local 當作 host，而是使用本機上的 IP 做為 host，原先不寫，預設是 localhost，但只限開發使用，如果要實際部屬，需額外設定是 0.0.0.0。

先使用以下指令，建立 Docker Image：

```
docker build -f Dockerfile -t api:7.2.2 .
```

建立好 Image 之後，以下是 Docker-Compose：

https://github.com/FinMind/FinMindBook/blob/master/DataEngineering/Chapter7/7.2/7.2.2/api.yml

```
api.yml
version: '3.0'
services:
  api:
    image: api:7.2.2
    ports:
      - 8888:8888
    hostname: "api"
    restart: always
    environment:
      - TZ=Asia/Taipei
```

```
    networks:
        - my_network

networks:
  my_network:
    # 加入已經存在的網路
    external: true
```

使用以下指令啟動 API

```
docker-compose -f api.yml up
```

接著如果看到以下畫面，就代表成功了。

```
sam@DESKTOP-IKT69L5:~/FinMindBook/DataEngineering/Chapter7/7.2/7.2.2$ docker-compose -f api.yml up
Creating 722_api_1 ... done
Attaching to 722_api_1
api_1  | Loading .env environment variables...
api_1  | INFO:     Started server process [7]
api_1  | INFO:     Waiting for application startup.
api_1  | INFO:     Application startup complete.
api_1  | INFO:     Uvicorn running on http://0.0.0.0:8888 (Press CTRL+C to quit)
```

打開 http://127.0.0.1:8888/docs 此連結，會看到以下畫面。

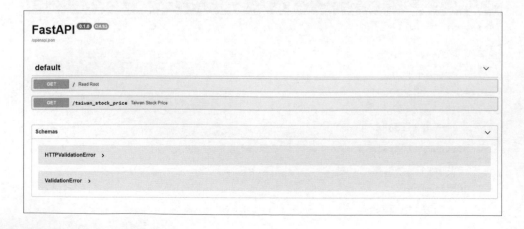

代表成功使用 Docker，架設好 API 了。

這時真實情境是，在開發機上做開發，並測試好用 Docker 架設 API，一切正常後，可以直接將 API Image，上傳到正式機上，並用 api.yml 啟動 Docker 部屬 API，並且不受開發機、正式機不同環境的限制了。

本書到此，成功使用 Docker 架設爬蟲、資料庫、RabbitMQ、API 以上服務了，同時也建立了專屬的爬蟲、API Docker Image，那要如何管理這些 Image 呢？以下將介紹 Docker Hub https://hub.docker.com/，Docker 官方網站，可以協助使用者，管理自己的 Image。

7.3 發布 Docker Image

7.3.1 Docker Hub 介紹

https://hub.docker.com/，Docker Hub，類似於 GitHub，GitHub 的 Repo 有 Public、Private 兩個方案，Public 是對外開源用途，Private 是私人開發用，Docker Hub 也一樣，有 Public、Private，例如想做一個開源服務，可以直接發布 Public 的 Docker Image，讓其他人直接下載，就像本書使用的 MySQL、RabbitMQ 這兩種服務，也是使用別人 Public 的 Image。

而 Private 可以存放不想公開的 Image，如爬蟲、API，在 local 開發好程式，並 Build 成 Image 後，上傳 Docker Hub，再到正式機，拉 Image 下來部屬。

7.3.2 發布 Docker Image

首先，請先到 Docker Hub 註冊，註冊後，先用 Docker 登入，（以下使用
者名稱，用 linsamtw 為例，你需要更改為自己的）。

```
docker login --username linsamtw
```

接著再輸入密碼，登入成功後，會顯示以下畫面：

```
sam@DESKTOP-IKT69L5:~/FinMindBook/DataEnginner/Chapter6/6.3/6.3.2$ docker login --username linsamtw
Password:
WARNING! Your password will be stored unencrypted in /home/sam/.docker/config.json.
Configure a credential helper to remove this warning. See
https://docs.docker.com/engine/reference/commandline/login/#credentials-store

Login Succeeded
```

上傳 Image 前，需要對 Image 下 tag，如下：

```
docker tag crawler:7.2.1 linsamtw/crawler:7.2.1
```

上傳爬蟲的 Docker Image：

```
docker push linsamtw/crawler:7.2.1
```

同樣，也上傳 API 的 Docker Image：

```
docker tag api:7.2.2 linsamtw/api:7.2.2
docker push linsamtw/api:7.2.2
```

這時上去 Docker Hub，就可以看到剛剛發布的 Image，以下是本書上傳
的 Image。

那在 Local Build 好 Image 後，要如何在雲端正式機部屬呢？

7.4 雲端部屬

先到雲端正式機，使用 docker login 指令登入後，使用 pull 指令，將發布的 Image 拉下來。

```
docker pull linsamtw/api:7.2.2
```

```
sam@localhost:~/FinMindBook/DataEngineering/Chapter7/7.2/7.2.2$ docker pull linsamtw/api:7.2.2
7.2.2: Pulling from linsamtw/api
8ad8b3f87b37: Already exists
66b9b7852117: Already exists
4823bea101d0: Already exists
76d893c0ed7f: Already exists
188b30775d2f: Pull complete
95688c554cf2: Pull complete
5a85af52400a: Pull complete
4f4fb700ef54: Pull complete
9728154404c8: Pull complete
481fd4d0b191: Pull complete
Digest: sha256:99ed3d3bf8a12a45a84c5d4d8241dd60067f71912e83e418c04bffbfffb4beee
Status: Downloaded newer image for linsamtw/api:7.2.2
docker.io/linsamtw/api:7.2.2
sam@localhost:~/FinMindBook/DataEngineering/Chapter7/7.2/7.2.2$ []
```

接下來，可以直接 Run 這個 Image，去部屬你的 API：

```
docker run -it --rm -p 8888:8888 linsamtw/api:7.2.2
```

由於啟動 API 指令，寫在 Dockerfile 裡，因此直接 Run Image，就部屬好
API 了。

```
sam@localhost:~/FinMindBook/DataEngineering/Chapter7/7.2/7.2.2$ docker run -it --rm -p 8888:8888 linsamtw/api:7.2.2
Loading .env environment variables...
INFO:     Started server process [7]
INFO:     Waiting for application startup.
INFO:     Application startup complete.
INFO:     Uvicorn running on http://0.0.0.0:8888 (Press CTRL+C to quit)
```

那要如何在背景執行呢？總不可能一直開著這個 Terminal，只需要在啟動
Docker Image 時，多加一個指令 -d 即可，如下：

```
docker run -d -it --rm -p 8888:8888 linsamtw/api:7.2.2
```

這時就在正式機，部屬好 API 了。

同樣的，在 7.2，使用 Docker 在 local 建立起 RabbitMQ、MySQL、爬
蟲，這時模仿部屬 API，在雲端部屬以上服務，就可以在雲端，真正地部
屬整個產品環境了，因此整個開發流程是：

1. Local 開發完，測試沒問題。
2. 在 Local Build 成 Image，Push 至 Docker Hub。
3. 至雲端機器 Pull Image，部屬。

這樣能確保程式上，不受環境影響，且也使用 config.py、local.ini，去控
管開發與正式環境的設定，這能讓開發上，輕易切換不同環境。

到目前為止，部屬了爬蟲、API、RabbitMQ、MySQL，4 種不同的服務，都用 Docker 啟動，那有沒有什麼好的管理工具，且能統一管理、更新、查看所有 Log 並做監控呢？總不可能去一台台雲端上，查看 Docker Log，要換版時，總不可能上去一台台的機器更換 Docker Image，如果有 10 台機器，那這樣也太笨了。

當然你可以自己寫程式做，但這問題我們不是第一個遇到，秉持著不要自己造輪子的想法，將介紹 Docker Swarm 這套架構。

可能有些讀者會疑問，為什麼不選用 K8s。網路上蠻多 Swarm vs K8s 的文章，Dwarm 主要是簡單、容易學習使用，K8s 多數用於大型架構，且大多不會自架，而是直接使用 GCP 等雲服務，專案剛起步，先選用 Swarm，相對來說，開發比較快速，且 GCP 也不便宜，大多都只有公司玩得起，還在學習階段，先用 Swarm 比較親民。

7.5 Docker Swarm

7.5.1 介紹

本書到現在，介紹了 Docker，用於啟動單一服務。Docker-Compose，管理多個 Docker。如果遇到多個 Docker-Compose 時，要如何管理呢？剛好在本書情境，就有 API、爬蟲、RabbitMQ、MySQL 這幾個 Docker-Compose，這時就要使用 Docker Swarm 來進行管理，如下圖。

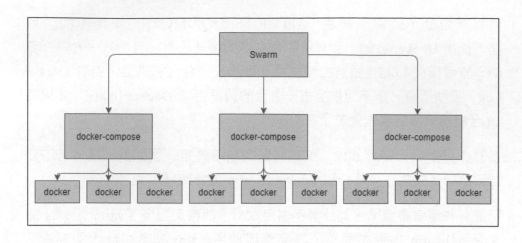

另外，Swarm 可以管理多台機器，這邊引入一個概念，**Manager** 和 **Worker**。在多台機器架構下，你會有一個負責作管控的 Manager 機器，剩下其他的機器都是 Worker，Manager 有 Swarm 的管理權限，例如部屬線上服務、更新線上服務、指定 Worker 執行特定任務等等，而 Worker 就只是單純做 Manager 分配好的工作，包含建立 MySQL、RabbitMQ、API、爬蟲等，如以下情境。

我有 A、B 兩台機器，Manager 在 A 機器上 Worker 在 B 機器上，要部屬 MySQL，那要怎麼做呢？

首先，在 Manager 下達指令，要部屬 MySQL 這個服務在 Swarm 中，並指定在 Worker B 上跑 MySQL 這個 Docker Container，這時 B 接收到指令，並會自己去抓取相對應的 MySQL Image，並啟動這個 Image。

流程大致上就是這樣，管理權限全部掌握在 Manager 手上，所有的事，你直接在 Manager 下指令即可，他會傳達給 Worker，而 Worker 只是工人，負責跑 Manager 決定的工作，基本上不會直接操作到 Worker。

實際狀況如下，將 MySQL、RabbitMQ、API、Crawle 分別跑在不同機器上。

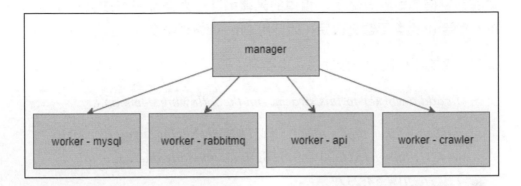

這樣的好處是，全部的 Services 都統一管理，相較於連去一台台的機器部屬 MySQL、RabbitMQ 等，甚至調整 MySQL、RabbitMQ 的 config 設定檔，我只要在 Manager 上，就可以管理所有的機器。

下個章節，將示範如何建立第一個 Swarm，並讓 Node 與 Manager 做連結。

7.5.2 建立第一個 swarm

首先，初始化 Docker Swarm：

```
docker swarm init
```

畫面如下：

```
sam@localhost:~$ docker swarm init
Swarm initialized: current node (w5m87thetvi8e8kwifzkxscfm) is now a manager.

To add a worker to this swarm, run the following command:

    docker swarm join --token SWMTKN-1-1loirfpq5uyajclzor35t1ws4xt1z4hhwh141fn23vg5ntr58g-0p85mm8epx3ba5c3qqrtdlfr1 139.162.104.54:2377

To add a manager to this swarm, run 'docker swarm join-token manager' and follow the instructions.

sam@localhost:~$
```

這邊要特別注意，139.162.104.54 是 Linode 的實體 IP，如果你用個人電腦架設 Swarm，使用虛擬 IP，那其他電腦是找不到你的 IP 位址的，一般家庭跟電信商如中華電信，租用的光纖網路，大多都只有虛擬（浮動）IP，實體 IP 還需要額外設定，因此用 Linode 方便很多。

接著注意初始化後的一段訊息，上面寫：

> To add a worker to this swarm, run the following command:
>
> docker swarm join --token SWMTKN-1-1loirfpq5uyajclzor35t1
> ws4xt1z4hhwh141fn23vg5ntr58g-0p85mm8epx3ba5c3qqrtdlfr1
> 139.162.104.54:2377
>
> To add a manager to this swarm, run 'docker swarm join-token
> manager' and follow the instructions.

也就是説，如果你要加一個 Worker 到 Swarm，使用以下指令：

```
docker swarm join --token SWMTKN-1-1loirfpq5uyajclzor35t1ws4xt1z4hhwh141
fn23vg5ntr58g-0p85mm8epx3ba5c3qqrtdlfr1 139.162.104.54:2377
```

你可能會問，這段指令太長，需要額外記下來嗎？不用，在 Manager 的機器上，使用以下指令，就會再次出現：

```
docker swarm join-token worker
```

```
sam@localhost:~$ docker swarm join-token worker
To add a worker to this swarm, run the following command:

    docker swarm join --token SWMTKN-1-1loirfpq5uyajclzor35t1ws4xt1z4hhwh141fn23vg5ntr58g-0p85mm8epx3ba5c3qqrtdlfr1 139.162.104.54:2377
```

這時就初始化成功了，只要在其他電腦上，執行以上指令，就會加入 Docker Wwarm 中，但沒有一個 UI 介面，怎麼知道目前 Swarm 下，有哪些服務或是 Worker 呢？下一章節，將介紹 Portainer，Swarm 的 UI 介面之一。

7.5.3 管理介面 --- Portainer

以下是 Portainer 的畫面，以 FinMind 為例，Swarm 內包含 57 個 Services、221 個 Containers，管理多種服務，所以 Swarm 的上限是很高的，以下將提供範例，一步步建立出你自己的 Portainer、Swarm 架構。

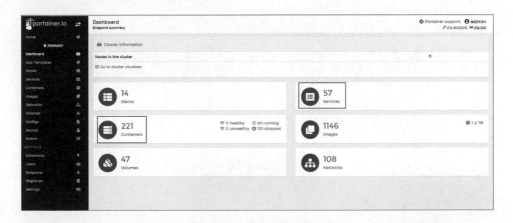

首先，一樣用 Docker 架設 Portainer，使用以下 Docker-Compose：

https://github.com/FinMind/FinMindBook/blob/master/DataEngineering/Chapter7/7.5/7.5.3/portainer.yml

```
portainer.yml
version: '3.2'

services:
  agent:
    image: portainer/agent
    environment:
```

```
      AGENT_CLUSTER_ADDR: tasks.agent
    volumes:
      - /var/run/docker.sock:/var/run/docker.sock
      - /var/lib/docker/volumes:/var/lib/docker/volumes
    networks:
      - agent_network
    deploy:
      mode: global
      placement:
        constraints: [node.platform.os == linux]

  portainer:
    image: portainer/portainer:1.24.2
    command: -H tcp://tasks.agent:9001 --tlsskipverify
    ports:
      - "9000:9000"
      - "8000:8000"
    volumes:
      - portainer_data:/data
    networks:
      - agent_network
    deploy:
      mode: replicated
      replicas: 1
      placement:
        constraints: [node.role == manager]

networks:
  agent_network:
    driver: overlay

volumes:
  portainer_data:
```

在 Swarm 架構下，不再是使用 Docker-Compose 啟動，而是改用以下指令：

```
docker stack deploy -c portainer.yml por
```

在解釋指令之前，先來看 Portainer 畫面，會比較好理解，進入以下連結，IP 需換成你自己 Linode 的 IP：

http://139.162.104.54:9000/

會看到以下畫面：

一開始，需要先設定帳號密碼，設定完後，按 Create user，之後會進入以下畫面，點選下面那隻鯨魚：

接著跳轉到以下畫面,再點選 Service:

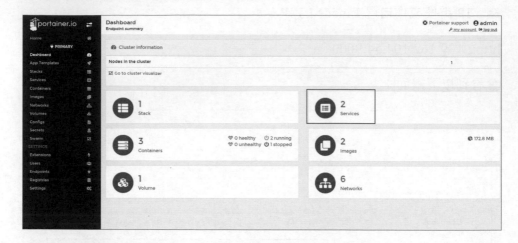

跳轉到此畫面:

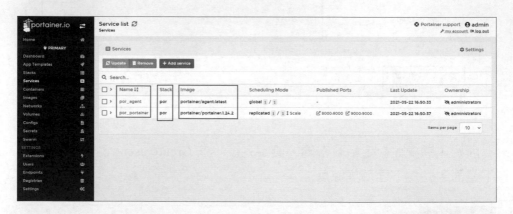

這時就出現剛剛用 portainer.yml 架設的 Container,先來介紹幾個名詞,
Name 底下的 por_agent、por_portainer,分別是上面 portainer.yml 中的兩個
services name,agent 和 portainer,如下,前面的 por,就是 Stack Name,

```
services:
  agent:
    image: portainer/agent
    environment:
      AGENT_CLUSTER_ADDR: tasks.agent
    volumes:
      - /var/run/docker.sock:/var/run/docker.sock
      - /var/lib/docker/volumes:/var/lib/docker/volumes
    networks:
      - agent_network
    deploy:
      mode: global
      placement:
        constraints: [node.platform.os == linux]

  portainer:
    image: portainer/portainer:1.24.2
    command: -H tcp://tasks.agent:9001 --tlsskipverify
    ports:
      - "9000:9000"
      - "8000:8000"
```

這時出現一個新名詞，Stack，英文翻譯是堆，如下圖，以目前範例來說，Stack 底下包含 agent、portainer，Stack 是一包、一堆的概念，例如 stack - por 這一堆，下面包含 agent、portainer 這兩個 Container，那為什麼要分 Stack 呢？因為後續包含更新服務、刪除服務等，都是以 Stack 為單位做操作。

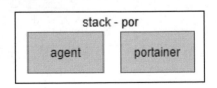

接下來，將以上用 Docker-Compose 部屬的服務，全部改到 Swarm 上。初期會花許多時間在架構上，因為目前服務少，所以感受不到好處，可能還會想說，不如直接用 Docker-Compose，但是考量到做 Data 專案，不

只是爬蟲、資料庫、API，還需要 Web 呈現視覺化、分析結果、紀錄 Log 等等，未來需要更多服務，因此用 Swarm 做管理，在未來拓展新服務時，可以直接在 Manager 操作，方便很多。

7.6 部屬服務

本章節，要部屬 MySQL、RabbitMQ、爬蟲、API 到 Swarm，架構如下：

```
├──── Makefile
├──── README.md
├──── api
├──── financialdata
├──── mysql.yml
└──── rabbitmq.yml
```

MySQL、RabbitMQ 分別使用 yml 就能部屬，api 跟 financialdata，需要建立 Docker Image，這兩個是獨立的 Project，建議都開新的 Project 做版控，FinMind 使用的是 GitLab，主要是因為要使用 GitLab-CI 進行後續的 CICD 自動化測試、部屬。

7.6.1 MySQL

將 MySQL 架設到 Swarm 架構中，跟 5.3.2 一樣，先在雲端機器上，建立 MySQL Volume。

```
docker volume create mysql
```

建立 Network

```
docker network create --scope=swarm --driver=overlay my_network
```

如果發生 *network with name my_network already exists* 的問題，請先
使用以下指令，刪除 my_network，如無法刪除，請先使用 docker stop
container_name 指令，停止相關 Container，再刪除 my_network。

```
docker network rm my_network
```

這裡跟之前比起來，比較特別，需要指定 **scope=swarm**，且 driver 指定
overlay，scope 指的就是 **network** 範圍，這裡用 **swarm**，而 driver 設為
overlay，是讓 swarm 內的服務可以互通，只要在 swarm 內部的機器，不
論 5 台 10 台，都能互相通信。

接著使用一樣的 mysql.yml，並加上 Swarm 設定，如下：

https://github.com/FinMind/FinMindBook/blob/master/DataEngineering/
Chapter7/7.6/mysql.yml

```
mysql.yml
version: '3'
services:

  mysql:
    image: mysql:8.0
    command: mysqld --default-authentication-plugin=mysql_native_password
    ports:
      - 3306:3306
    environment:
      MYSQL_DATABASE: mydb
      MYSQL_USER: user
      MYSQL_PASSWORD: test
```

```
            MYSQL_ROOT_PASSWORD: test
      volumes:
          - mysql:/var/lib/mysql
      # swarm 設定
      deploy:
        mode: replicated
        replicas: 1
        placement:
          constraints: [node.labels.mysql == true]
      networks:
          - my_network

phpmyadmin:
    image: phpmyadmin/phpmyadmin:5.1.0
    links:
        - mysql:db
    ports:
        - 8080:80
    environment:
        MYSQL_USER: user
        MYSQL_PASSWORD: test
        MYSQL_ROOT_PASSWORD: test
        PMA_HOST: mysql
    depends_on:
      - mysql
    # swarm 設定
    deploy:
      mode: replicated
      replicas: 1
      placement:
        constraints: [node.labels.mysql == true]
    networks:
        - my_network
```

```
networks:
  my_network:
    # 加入已經存在的網路
    external: true

volumes:
  mysql:
    external: true
```

這裡有兩點需要特別注意：

第一，**PMA_HOST**

```
environment:
    MYSQL_USER: user
    MYSQL_PASSWORD: test
    MYSQL_ROOT_PASSWORD: test
    PMA_HOST: mysql
```

在 phpmyadmin 要額外設定 environment，特別是 PMA_HOST，在 Swarm 架構下，需要指定 MySQL 的 host 位址，這也是跟 Docker-Compose 不太一樣的地方，一開始如果直接轉到 Swarm 下，會遇到許多坑，本書已經踩過，並提供解法。

第二，**labels**

```
constraints: [node.labels.mysql == true]
```

這是指定，要在特定 node 上跑 MySQL Services，Node 就等同於 Worker，例如我有 3 台機器，要怎麼指定特定服務如 MySQL，在不同機器上執行，就是靠這裡的設定。

在部屬 MySQL 之前，需要先到 Portainer 上做設定 Node 的 Labels，步驟如下：

1. 到 Portainer 頁面，點選 **Swarm**。

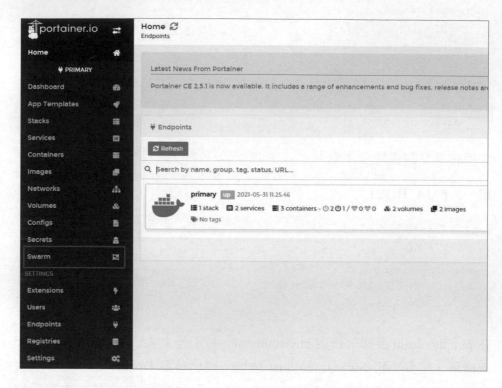

2. 點選 Nodes 底下的 **test**，這裡我是指定 test 這個 Worker 來架設 MySQL，如果你有多台機器，可以選其他機器。

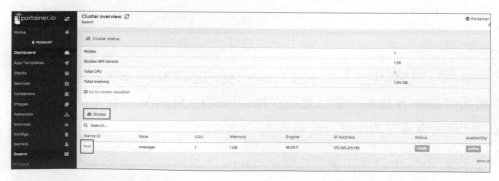

3. 接著點選 **Label**，會新增一個 Label，在 **Name** 的地方，輸入 **MySQL**，在 **Value** 的地方，輸入 **true**，最後點選 **Apply changes**，就成功加上 Label 了。

4. 有 Label 之後，使用以下指令，這是專門用來部屬 Swarm 的指令。

```
docker stack deploy --with-registry-auth -c mysql.yml mysql
```

5. 這時候在 Swarm 上，就可以看到 MySQL 的 Services。

6. preparing 是這個 Services 正在準備中的狀態，等 Docker 成功 Run 起
來後，會改成以下狀態，**running**。

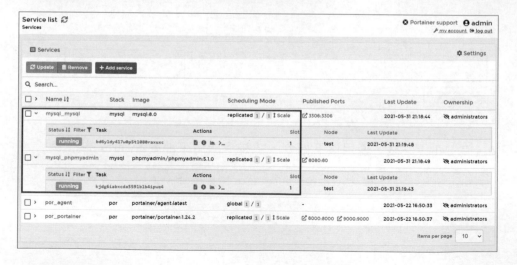

這時成功部屬好 MySQL，到 http://139.162.104.54:8080/（139.162.104.54
改成你機器的 IP），就可以看到 MySQL 了。

使用 Portainer 介面，就可以很清楚的知道，MySQL 的狀態，接著部屬
RabbitMQ。

7.6.2 RabbitMQ

使用的 Docker-Compose 如下，同樣需要做 Swarm 設定：

https://github.com/FinMind/FinMindBook/blob/master/DataEngineering/
Chapter7/7.6/rabbitmq.yml

rabbitmq.yml

```yaml
version: '3'
services:

  rabbitmq:
    image: 'rabbitmq:3.6-management-alpine'
    ports:
      - '5672:5672'
      - '15672:15672'
    environment:
      RABBITMQ_DEFAULT_USER: "worker"
      RABBITMQ_DEFAULT_PASS: "worker"
      RABBITMQ_DEFAULT_VHOST: "/"
    # swarm 設定
    deploy:
      mode: replicated
      replicas: 1
      placement:
        constraints: [node.labels.rabbitmq == true]
    networks:
      - my_network

  flower:
    image: mher/flower:0.9.5
    command: ["flower", "--broker=amqp://worker:worker@rabbitmq",
"--port=5555"]
    ports:
      - 5555:5555
    depends_on:
      - rabbitmq
    # swarm 設定
```

```
deploy:
  mode: replicated
  replicas: 1
  placement:
    constraints: [node.labels.flower == true]
  networks:
    - my_network


networks:
  my_network:
    # 加入已經存在的網路
    external: true
```

同樣，需要先到 Portainer 上，將 Labels 加上去，如下：

接著執行以下指令：

```
docker stack deploy --with-registry-auth -c rabbitmq.yml rabbitmq
```

這時在 Portainer 上，就能看到 RabbitMQ、Flower 這兩個 Services，如
下：

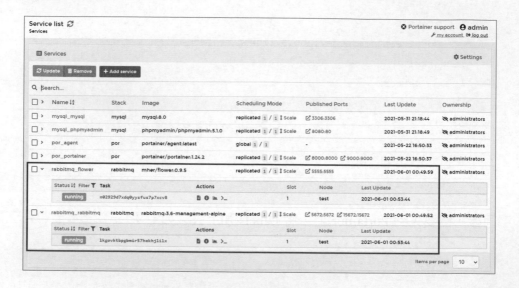

這時到 http://139.162.104.54:15672/（139.162.104.54 改成你機器的 IP），
就會看到 RabbitMQ 的畫面：

到 http://139.162.104.54:5555/（139.162.104.54 改成你機器的 IP），就會
看到 Flower 的畫面。

如果畫面都跟上面一樣,那就代表部屬成功了,本書到此,已經使用 Docker Swarm,同時管理 MySQL、Flower、RabbitMQ,下一步,把爬蟲也部屬上去。

7.6.3 爬蟲

首先,到雲端機器上,將章節 7.3.2 上傳 Docker Hub 的 Docker Image, Pull 下來:

```
docker pull linsamtw/crawler:7.2.1
```

當然你也可以自己重新 Build,指令就參考前面章節,接著使用以下 Docker-Compose 部屬在 Swarm 上:

https://github.com/FinMind/FinMindBook/blob/master/DataEngineering/ Chapter7/7.6/financialdata/crawler.yml

crawler.yml
```
version: '3.0'
services:
  crawler_twse:
    image: linsamtw/crawler:7.2.1
```

```
    hostname: "twse"
    command: pipenv run celery -A financialdata.tasks.worker worker
--loglevel=info --concurrency=1  --hostname=%h -Q twse
    restart: always
    # swarm 設定
    deploy:
      mode: replicated
      replicas: 1
      placement:
        constraints: [node.labels.crawler_twse == true]
      environment:
      - TZ=Asia/Taipei
      networks:
        - my_network

networks:
  my_network:
    # 加入已經存在的網路
    external: true
```

https://github.com/FinMind/FinMindBook/blob/master/DataEngineering/
Chapter7/7.6/financialdata/scheduler.yml

scheduler.yml
```
version: '3.0'
services:
  scheduler:
    image: linsamtw/crawler:7.2.1
    hostname: "twse"
    command: pipenv run python financialdata/scheduler.py
    restart: always
    # swarm 設定
```

```
  deploy:
    mode: replicated
    replicas: 1
    placement:
      constraints: [node.labels.crawler_scheduler == true]
    environment:
      - TZ=Asia/Taipei
    networks:
        - my_network

networks:
  my_network:
    # 加入已經存在的網路
    external: true
```

這裡同時部屬爬蟲跟 scheduler，跟上面相同，需要再次設定 Labels：

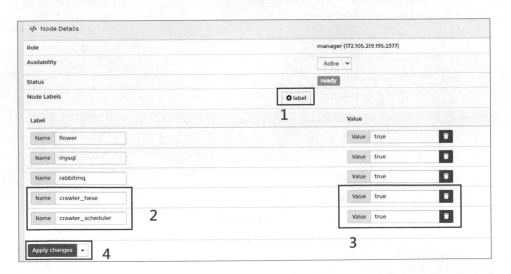

先部屬 crawler_twse 就好，之後很容易可以再拓展其他 Celery。因為 MySQL 是全新的，所以在部屬之前，需要先建立 DataBase，不然待會 Celery 的 Client，會連接不到 DataBase 報 Error。

將以下 SQL 複製到 MySQL 上並執行（此 SQL 包含建立 **financialdata** 資料庫，因此不需再額外手動建立 **financialdata** DataBase）。

https://github.com/FinMind/FinMindBook/blob/master/DataEngineering/ Chapter7/7.6/financialdata/create_partition_table.sql

```
create_partition_table.sql
CREATE DATABASE `financialdata`;
CREATE TABLE `financialdata`.`TaiwanStockPrice`(
    `StockID` VARCHAR(10) NOT NULL,
    `TradeVolume` BIGINT NOT NULL,
    `Transaction` INT NOT NULL,
    `TradeValue` BIGINT NOT NULL,
    `Open` FLOAT NOT NULL,
    `Max` FLOAT NOT NULL,
    `Min` FLOAT NOT NULL,
    `Close` FLOAT NOT NULL,
    `Change` FLOAT NOT NULL,
    `Date` DATE NOT NULL,
    PRIMARY KEY(`StockID`, `Date`)
)
PARTITION BY RANGE(YEAR(Date)) (
    PARTITION p2005 VALUES LESS THAN (2006),
    PARTITION p2006 VALUES LESS THAN (2007),
    PARTITION p2007 VALUES LESS THAN (2008),
    PARTITION p2008 VALUES LESS THAN (2009),
    PARTITION p2009 VALUES LESS THAN (2010),
```

```
        PARTITION p2010 VALUES LESS THAN (2011),
        PARTITION p2011 VALUES LESS THAN (2012),
        PARTITION p2012 VALUES LESS THAN (2013),
        PARTITION p2013 VALUES LESS THAN (2014),
        PARTITION p2014 VALUES LESS THAN (2015),
        PARTITION p2015 VALUES LESS THAN (2016),
        PARTITION p2016 VALUES LESS THAN (2017),
        PARTITION p2017 VALUES LESS THAN (2018),
        PARTITION p2018 VALUES LESS THAN (2019),
        PARTITION p2019 VALUES LESS THAN (2020),
        PARTITION p2020 VALUES LESS THAN (2021),
        PARTITION p2021 VALUES LESS THAN (2022),
        PARTITION p2022 VALUES LESS THAN (2023),
        PARTITION p2023 VALUES LESS THAN (2024),
        PARTITION p2024 VALUES LESS THAN (2025)

);

CREATE TABLE `financialdata`.`TaiwanFuturesDaily`(
    `Date` DATE NOT NULL,
    `FuturesID` VARCHAR(10) NOT NULL,
    `ContractDate` VARCHAR(30) NOT NULL,
    `Open` FLOAT NOT NULL,
    `Max` FLOAT NOT NULL,
    `Min` FLOAT NOT NULL,
    `Close` FLOAT NOT NULL,
    `Change` FLOAT NOT NULL,
    `ChangePer` FLOAT NOT NULL,
    `Volume` FLOAT NOT NULL,
    `SettlementPrice` FLOAT NOT NULL,
    `OpenInterest` INT NOT NULL,
```

```
    `TradingSession` VARCHAR(11) NOT NULL,
    PRIMARY KEY(`FuturesID`, `Date`)
)
PARTITION BY RANGE(YEAR(Date)) (
    PARTITION p2005 VALUES LESS THAN (2006),
    PARTITION p2006 VALUES LESS THAN (2007),
    PARTITION p2007 VALUES LESS THAN (2008),
    PARTITION p2008 VALUES LESS THAN (2009),
    PARTITION p2009 VALUES LESS THAN (2010),
    PARTITION p2010 VALUES LESS THAN (2011),
    PARTITION p2011 VALUES LESS THAN (2012),
    PARTITION p2012 VALUES LESS THAN (2013),
    PARTITION p2013 VALUES LESS THAN (2014),
    PARTITION p2014 VALUES LESS THAN (2015),
    PARTITION p2015 VALUES LESS THAN (2016),
    PARTITION p2016 VALUES LESS THAN (2017),
    PARTITION p2017 VALUES LESS THAN (2018),
    PARTITION p2018 VALUES LESS THAN (2019),
    PARTITION p2019 VALUES LESS THAN (2020),
    PARTITION p2020 VALUES LESS THAN (2021),
    PARTITION p2021 VALUES LESS THAN (2022),
    PARTITION p2022 VALUES LESS THAN (2023),
    PARTITION p2023 VALUES LESS THAN (2024),
    PARTITION p2024 VALUES LESS THAN (2025)
);
```

複製到以下地方：

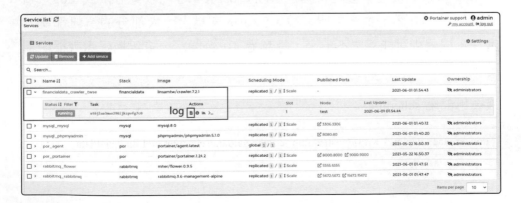

接著，使用以下指令做部屬爬蟲：

```
docker stack deploy --with-registry-auth -c financialdata/crawler.yml
financialdata
```

這時到 Portainer，就會出現 financialdata_crawler_twse，代表成功部屬，
如果想看 Celery log，可以直接點下面 Actions 第一個圖示：

跳轉到 Log 頁面後，Log 就顯示在以下畫面：

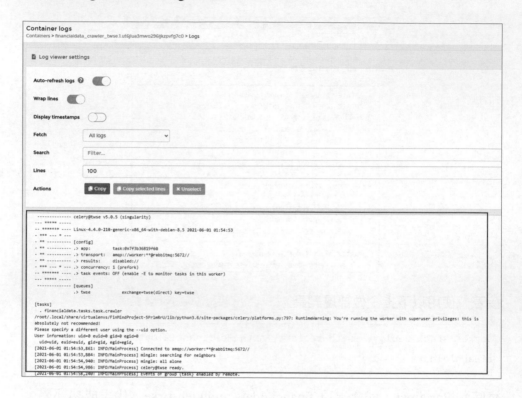

這是另一個好處，不用再去每台機器上收集、觀看 Log，Portainer 上面就有 Log 可以直接看，甚至有其他服務如 ELK，專門收集分散式 Log，這裡就不多作介紹，主要是想表達，Swarm 架構上，大部分的資訊都統一處理。機器少可能感覺不出來，等超過 10 台機器，這架構好處就很明顯了。

最後再部屬 scheduler：

```
docker stack deploy --with-registry-auth -c financialdata/scheduler.yml financialdata
```

執行以上指令，成功後，Portainer 上就會出現 financialdata_scheduler 的
Services 了。

到這步驟後，完成爬蟲部屬，並且有排程會每天抓新資料了，那歷史資
料怎麼辦呢？

點選下圖的按鈕：

再點選 Connect：

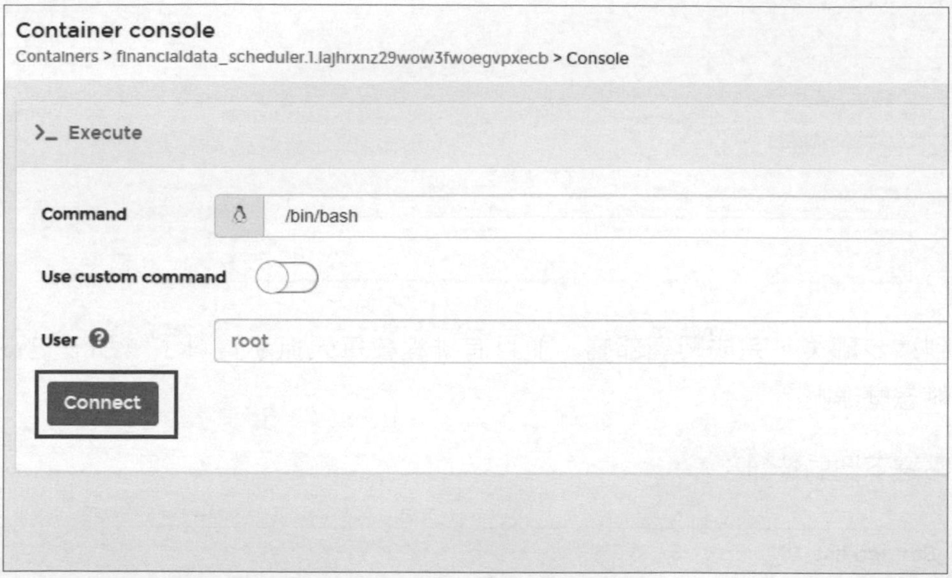

Container console
Containers > financialdata_scheduler.1.lajhrxnz29wow3fwoegvpxecb > Console

>_ Execute

| Command | ⚙ /bin/bash |

Use custom command ⬜

| User ❓ | root |

Connect

就會進入到 Container 內部：

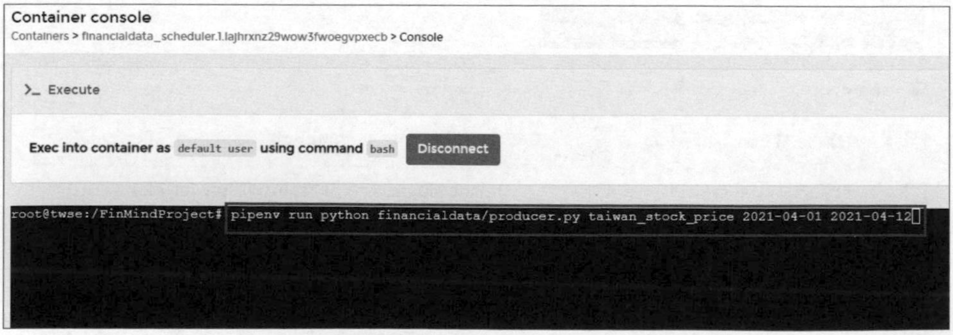

Container console
Containers > financialdata_scheduler.1.lajhrxnz29wow3fwoegvpxecb > Console

>_ Execute

Exec into container as default user using command bash Disconnect

```
root@twse:/FinMindProject# pipenv run python financialdata/producer.py taiwan_stock_price 2021-04-01 2021-04-12
```

這時就可以發送任務，抓歷史資料（日期可自行調整，這裡用 2021-04-01 到 2021-04-12 作為例子），指令如下：

```
pipenv run python financialdata/producer.py taiwan_stock_price 2021-04-01 2021-04-12
```

再去 RabbitMQ http://139.162.104.54:15672/#/queues（139.162.104.54 改成你機器的 IP）上查看 Queues 的狀態：

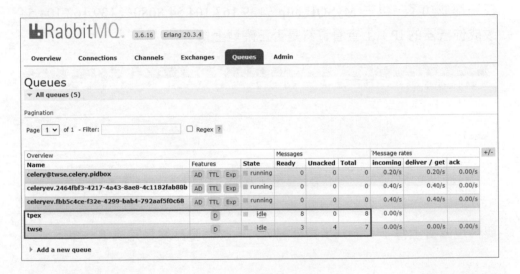

任務已經成功發送給 RabbitMQ，並且 twse 的 queue 已經被 Worker 執行了，但是 tpex 沒有，因為上面 Celery 只有啟動 twse 而已，這部分讀者可再自行調整。

那 Worker 執行任務，有沒有成功呢？到 http://139.162.104.54:5555/
（139.162.104.54 改成你機器的 IP）查看 Flower：

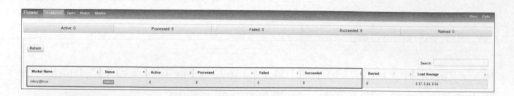

任務都成功了，再去 MySQL http://139.162.104.54:8080/（139.162.104.54
改成你機器的 IP）上查看資料是否正確地上傳：

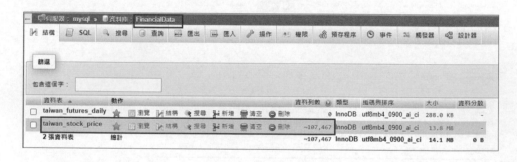

資料也確實上傳了，下一步再部屬 API，整個產品就算完成第一步了。

7.6.4　API

部屬 API Services，同樣，以下是 yml 檔：

https://github.com/FinMind/FinMindBook/blob/master/DataEngineering/
Chapter7/7.6/api/api.yml

```
api.yml
version: '3.0'
services:
  api:
    image: linsamtw/api:7.2.2
    ports:
        - 8888:8888
    hostname: "api"
    restart: always
    # swarm 設定
    deploy:
      mode: replicated
      replicas: 1
      placement:
        constraints: [node.labels.api == true]
    environment:
      - TZ=Asia/Taipei
    networks:
        - my_network

networks:
  my_network:
    # 加入已經存在的網路
    external: true
```

設定 deploy，labels 設定 api=true，接著去 Portainer 上面，設定要執行 API 機器的 Label，這裡依然使用同一台機器。

Docker 非常省資源，一台 5 美金的 Linode，就能同時部屬目前用到的所有服務：

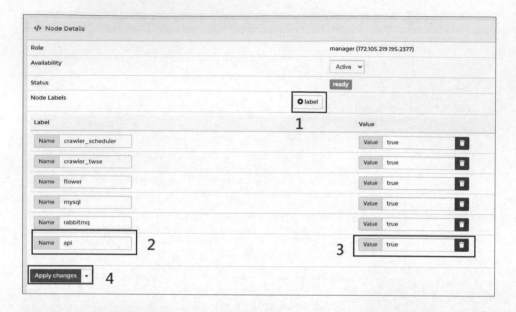

使用以下指令部屬 API：

```
docker stack deploy --with-registry-auth -c api/api.yml api
```

接著再到 Portainer 上看，就會出現 API 了。

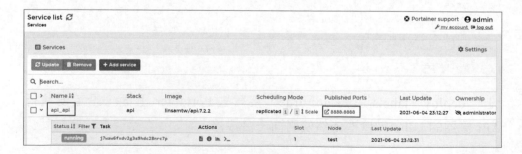

再到以下網址 http://139.162.104.54:8888/docs（139.162.104.54 改成你機器的 IP），就會看到成功部屬 API 了。

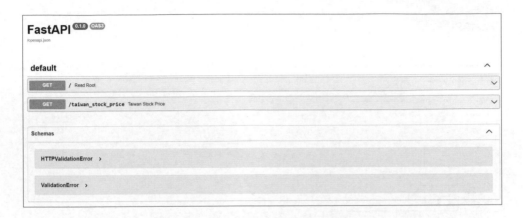

在 7.6.3，有成功爬蟲抓台股股價到資料庫，那試著對 API 發送 request，觀察是否拿的到資料，步驟如下：

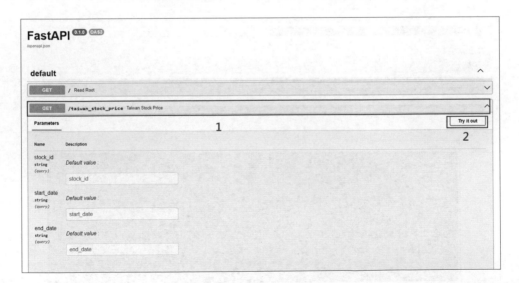

發送 API，前面爬取到 2021-04-01 的資料，因此這裡 start_date、end_date 的區間選 2021-04-01、2021-04-10：

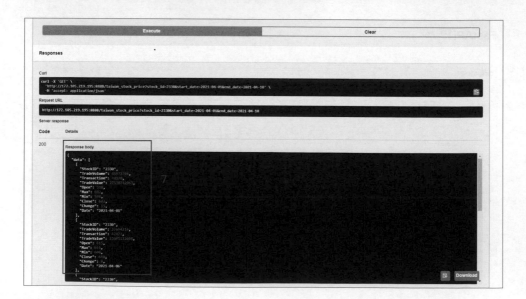

成功發送 API，且回傳 2330 股價資料，代表 API 成功。

本書到這，產品已經完成初版了，雖然目前只有台股股價，但整個流程已經串通，接下來只要照著架構持續開發新爬蟲，新 API，基本上不會有太大問題，且本書使用 Celery、RabbitMQ 分散式架構，因此效能問題不大，也不會被 ban IP，且雲端不怕機器壞掉，重點是，到現在，只使用 1~2 台機器而已，一個月只要 5~10 美金。

未來持續開發，架構慢慢變大，就會遇到幾個問題，**架構零散，導致重構困難，手動換版、手動測試，沒有固定流程，難以維護**，這也是團隊初期，最容易遇到的問題，急著開發新功能，導致欠缺測試、沒有 CICD 流程，造成換版常常掛掉，想重構卻容易把程式搞壞，難以維護等等的技術債，甚至最後砍掉重練（當然這是需要快速開發產品，做的取捨，本書單純探討工程方面的問題）。

為 了 解 決 以 上 問 題， 在 **Chapter 8、Chapter 9**， 將 介 紹 **Unit Test、CICD**，讓產品在開發一個階段，確定方向後，開始建立良好的開發流程，雖然會花許多時間，但可以保證後續產品的穩定性，不用擔心程式有 bug、新版本掛掉等問題。

某些讀者可能會認為，程式有 bug，這是開發者的問題，但筆者認為，這一部分也是管理者的責任，如果先把自動化測試做上去，就能夠自動把關，減少人為因素，當只管理一個專案，你可能有時間手動做測試，但如果同時管理 3 個以上的專案呢？自動化可以省去很多時間，減少錯誤發生，何樂而不為呢？

第 2 篇
產品迭代 -- 測試運維

自動化測試

8.1　單元測試 Unit Test

8.1.1　什麼是單元測試 Unit Test？

在程式開發中，會將許多程式寫成 function 函數，讓程式模組化，但在長期開發中，一定會經歷過**重構**階段，重新設計架構，那這時會遇到一個問題，要如何確保，重構完後，程式結果不會出錯呢？通常對程式做大改，很容易出現錯誤，這時就要引入測試，確保程式結果不變。

程式架構中，函數是最小單元，而單元測試，就是最小化的測試。將所有的函數，都做一個對應的測試函數，確保每一個函數不會因為某次改動而壞掉。除此之外，測試在多人開發上，有一個很大的幫助，其他人在看你寫的程式時，不需要直接執行程式，單純從測試上，就能清楚知道每個函數的 intput、output。

口頭上說可能無法理解，以下舉幾個例子作為範例。

8.1.2　第一個測試

首先，介紹第一個測試，讓讀者做個最基本的了解。關於 Unit Test 測試架構，有多種框架，如 Python 原生支援的 **unittest**、第三方框架 **pytest**，本書選用 pytest，主因是，pytest 支持 unittest、且在 GitHub 得到 7.4k stars，算是主流架構。

測試範例架構如下。

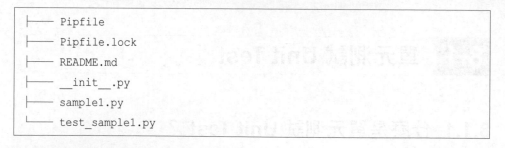

```
├── Pipfile
├── Pipfile.lock
├── README.md
├── __init__.py
├── sample1.py
└── test_sample1.py
```

安裝 pytest

```
pipenv install pytest==6.2.4
```

sample1.py 就是主程式，或是可以看成你的函數，**test_sample1.py** 就是對應的測試，基本上測試的程式規範是，在要測試的 .py 前面，加上 **test**，

例如 **sample1.py**，那對應的測試，就命名為 **test_sample1.py**，而程式碼如下。

https://github.com/FinMind/FinMindBook/blob/master/DataEngineering/
Chapter8/8.1.2/sample1.py

```
sample1.py

def add(x, y):
    return x + y
```

https://github.com/FinMind/FinMindBook/blob/master/DataEngineering/
Chapter8/8.1.2/test_sample1.py

```
test_sample1.py

from sample1 import add


def test_add():
    result = add(1, 2)
    expected = 3
    assert result == expected
```

舉個最簡單的例子，在 **sample1.py** 寫一個 **add** 函數，做加總，而測試就是，執行 **add** 函數。命名規則是，原函數前面加上 **test**，如 **test_add**，然後再寫上 **expected**，預期執行 **add** 函數得到的結果，最後設定 **assert**，檢查 **add** 函數結果，是否如預期。

接著要如何執行測試呢？

有兩種方法，使用 VS Code 插鍵，或是用 Command Line 指令執行，以下將一一介紹這兩種方法。

VS Code 插鍵

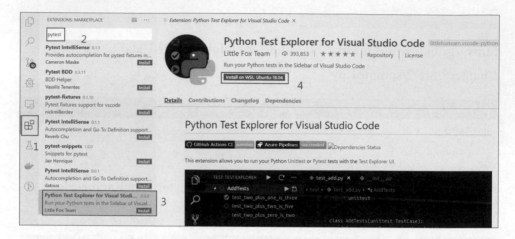

在 VS Code，根據以上步驟，安裝好 pytest 插鍵後，重啟 VS Code，之後
會看到以下畫面：

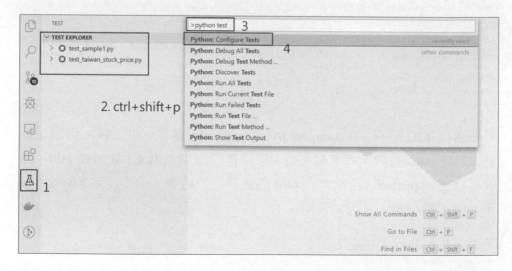

你的 VS Code IDE，左邊的欄位，會多一個藥水，這是 VS Code 提供的
測試工具，但還沒有任何的測試，需要設定 VS Code，選用哪個 Python
的測試框架。

鍵盤輸入 **ctrl+shift+p**，搜尋 **python test**，選取 **Python: Configure**：

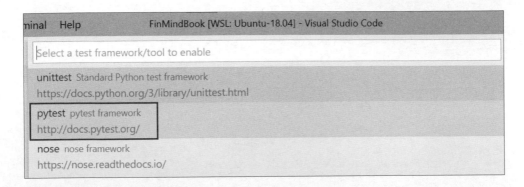

VS Code 會出現三個測試框架，選擇 pytest，接著 VS Code 需要選擇，要測試哪個目錄路徑，選取第一個選項．，當前路徑：

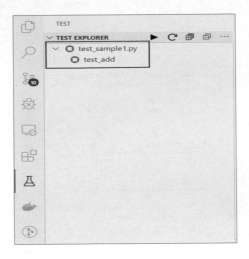

設定完成，左邊畫面出現了測試：

接著本書將介紹，如何使用 GUI 工具，執行測試。

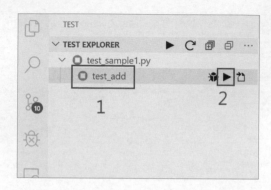

在測試的地方，可以看到，剛剛寫的 test_sample1.py，而底下的 test_add，就是測試函數，接著點擊箭頭（2），就會開始執行 test_add 的測試。

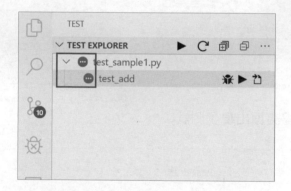

藍色的 …，代表正在執行中測試，成功就會顯示綠色 ✓。

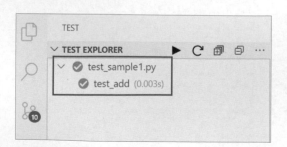

VS Code 插件，主要是在開發上，方便測試單一函數，在專案開發中，你
可能有上百個測試，跑一次通常都需要花費不少時間，因此只測特定函
數，是非常有幫助的。接下來介紹，使用 Command line 指令做測試。

command line

使用以下指令做測試：

```
pipenv run pytest test_sample1.py
```

```
sam@DESKTOP-IKT69LS:~/FinMindBook/DataEngineering/Chapter7/7.2/7.2.1$ docker-compose -f scheduler.yml up
WARNING: Found orphan containers (721_phpmyadmin_1, 721_crawler_twse_1, 721_rabbitmq_1, 721_mysql_1, 721_flower_1) for this project
. If you removed or renamed this service in your compose file, you can run this command with the --remove-orphans flag to clean it
up.
Creating 721_scheduler_1 ... done
Attaching to 721_scheduler_1
scheduler_1  | Loading .env environment variables...
scheduler_1  | 2021-10-31 00:45:46.504 | INFO     | __main__:main:27 - sent_crawler_task
```

成功後，就會顯示上圖，這就代表測試成功了，那指令有什麼好處呢？
在 Chapter 9，將介紹 CI 自動化測試，寫流程自動跑測試，一般在 Git 開
發上，每次 Merge 前，都必須跑過測試，過了才能安心 Merge，因此需
要指令做測試，而在 VS Code 上手動跑測試，不符合自動化的概念，VS
Code 測試主要是開發上使用。

介紹完簡單的測試之後，將使用前面開發的爬蟲、API 專案做範例，介紹
實務上的測試怎麼執行。

8.1.3 測試覆蓋率

在介紹爬蟲、API 測試範例之前，先介紹一個測試的進階概念，**測試覆
蓋率（Test Coverage）**。顧名思義，這是判斷整個專案，被測試覆蓋的比
例，舉例來說，你的專案有 10 個 function 函數，而你只寫了 3 個對應的
測試，那你的覆蓋率就是，3/10，30%，代表你的測試還有很大的努力空
間。

測試覆蓋率，當然是越高越好，但這邊要先提醒一件事，即使到了 100%
覆蓋率，依然不代表程式不會出錯，很多情況，是商業邏輯上的判斷問
題，而非程式本身的問題。

最後，測試不是萬能，但是沒有測試，是萬萬不能，覆蓋率越高，越能
保證你的程式穩定度。

8.1.4　爬蟲單元測試範例

爬蟲也可以寫測試嗎？筆者過去聽過一種說法，爬蟲因為是去抓取對方
網站，而對方網站會不穩定，所以寫爬蟲測試沒用。

這其實是一個不好的觀念，因為測試無法 100% 保證程式穩定性，所以就
不做嗎？不做就是 0 分，做了至少有 60 分，以下介紹爬蟲測試範例。

架構如下，tests 資料夾與 financialdata 在同一層級，tests/crawler/test_
taiwan_stock_price.py，這隻程式就是台股爬蟲的測試。

https://github.com/FinMind/FinMindBook/tree/master/DataEngineering/Chapter8/8.1.4

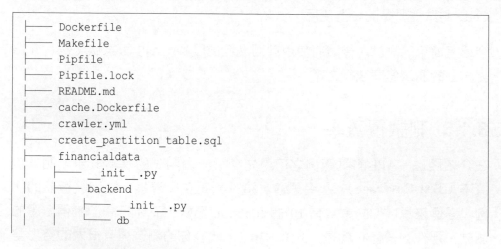

```
├──   Dockerfile
├──   Makefile
├──   Pipfile
├──   Pipfile.lock
├──   README.md
├──   cache.Dockerfile
├──   crawler.yml
├──   create_partition_table.sql
├──   financialdata
│     ├──   __init__.py
│     ├──   backend
│     │     ├──   __init__.py
│     │     └──   db
```

```
│   │           ├── __init__.py
│   │           ├── clients.py
│   │           ├── db.py
│   │           └── router.py
│   ├── config.py
│   ├── crawler
│   │   ├── __init__.py
│   │   ├── taiwan_futures_daily.py
│   │   └── taiwan_stock_price.py
│   ├── producer.py
│   ├── scheduler.py
│   ├── schema
│   │   ├── __init__.py
│   │   └── dataset.py
│   └── tasks
│       ├── __init__.py
│       ├── task.py
│       └── worker.py
├── genenv.py
├── local.ini
├── pytest.ini
├── .coveragerc
├── scheduler.yml
├── setup.py
└── tests
    ├── __init__.py
    └── crawler
        ├── __init__.py
        └── test_taiwan_stock_price.py
```

這裡新增了 pytest.ini、.coveragerc 這兩個關於測試的設定檔,主要是測試設定檔、測試覆蓋率設定檔,另外新增了兩個 Package,用於進階的測試與測試覆蓋率計算,安裝方法如下:

```
pipenv install pytest-cov==2.11.1 pytest-mock==3.5.1
```

安裝完之後,來看看測試範例,test_taiwan_stock_price.py 的覆蓋率,指令如下:

```
pipenv run pytest --cov-report term-missing --cov-config=.coveragerc
--cov=./financialdata/ tests/
```

測試結果如下圖：

框框 1，Cover，是每個主程式的測試覆蓋率，這個指令會自動計算覆蓋率，這裡先以 taiwan_stock_price.py 的測試作範例，因此其他的 Cover 都是 0。

框框 2，這裡 test_taiwan_stock_price.py 測試覆蓋率達到 100%，代表爬蟲程式中的各種情境，都被測試覆蓋到。

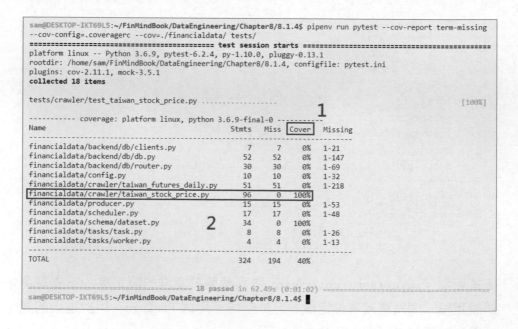

以上測試過了，測試率也到達 100%，接下來，將一步步介紹測試程式碼。

這裡需特別注意一點，主程式 taiwan_stock_price.py 只有 316 列，但 test_taiwan_stock_price.py 卻有 581 列，這也是為什麼大部分的人，不喜歡也缺少寫測試的習慣，因為在 coding 上，會花掉至少一半的開發時間。

由於專案初期開發上，會要求產品快速產出，因此缺少測試，是很合理的。但產品到一定階段後，必須將測試補上，不然整份專案將變成一塊磚，沒有人敢去動，因為非常容易把程式搞壞，後續的人更難維護。

https://github.com/FinMind/FinMindBook/blob/master/DataEngineering/
Chapter8/8.1.4/tests/crawler/test_taiwan_stock_price.py

test_taiwan_stock_price.py，關於每個測試的情境、細節，本書都寫在註解上：

```python
import pandas as pd

from financialdata.crawler.taiwan_stock_price import (
    clear_data,
    colname_zh2en,
    convert_change,
    convert_date,
    crawler,
    gen_task_paramter_list,
    is_weekend,
    set_column,
    twse_header,
    tpex_header,
    crawler_twse,
    crawler_tpex,
)
from financialdata.schema.dataset import (
    check_schema,
)


def test_is_weekend_false():
```

```python
    """
    測試, 非周末, 輸入周一 1, 回傳 False
    """
    result = is_weekend(day=1)  # 執行結果
    expected = False
    # 先寫好預期結果, 這樣即使不執行程式,
    # 單純看測試, 也能了解這個程式的執行結果
    assert (
        result == expected
    )  # 檢查, 執行結果 == 預期結果


def test_is_weekend_true():
    """
    測試, 是周末, 輸入週日 0, 回傳 False
    """
    result = is_weekend(day=0)  # 執行結果
    expected = True
    # 先寫好預期結果, 這樣即使不執行程式,
    # 單純看測試, 也能了解這個程式的執行結果
    assert (
        result == expected
    )  # 檢查, 執行結果 == 預期結果


def test_gen_task_paramter_list():
    """
    測試建立 task 參數列表, 2021-01-01 ~ 2021-01-05
    """
    result = gen_task_paramter_list(
        start_date="2021-01-01",
        end_date="2021-01-05",
```

```
)   # 執行結果
expected = [
    {
        "date": "2021-01-01",
        "data_source": "twse",
    },
    {
        "date": "2021-01-01",
        "data_source": "tpex",
    },
    {
        "date": "2021-01-02",
        "data_source": "twse",
    },
    {
        "date": "2021-01-02",
        "data_source": "tpex",
    },
    {
        "date": "2021-01-05",
        "data_source": "twse",
    },
    {
        "date": "2021-01-05",
        "data_source": "tpex",
    },
]
# 預期得到 2021-01-01 ~ 2021-01-05 的任務參數列表
# 再發送這些參數到 rabbitmq, 給每個 worker 單獨執行爬蟲
assert (
    result == expected
)   # 檢查, 執行結果 == 預期結果
```

```python
def test_clear_data():
    # 準備好 input 的假資料
    df = pd.DataFrame(
        [
            {
                "StockID": "0050",
                "TradeVolume": "4,962,514",
                "Transaction": "6,179",
                "TradeValue": "616,480,760",
                "Open": "124.20",
                "Max": "124.65",
                "Min": "123.75",
                "Close": "124.60",
                "Change": 0.25,
                "Date": "2021-01-05",
            },
            {
                "StockID": "0051",
                "TradeVolume": "175,269",
                "Transaction": "44",
                "TradeValue": "7,827,387",
                "Open": "44.60",
                "Max": "44.74",
                "Min": "44.39",
                "Close": "44.64",
                "Change": 0.04,
                "Date": "2021-01-05",
            },
            {
                "StockID": "0052",
```

```
                "TradeVolume": "1,536,598",
                "Transaction": "673",
                "TradeValue": "172,232,526",
                "Open": "112.10",
                "Max": "112.90",
                "Min": "111.15",
                "Close": "112.90",
                "Change": 0.8,
                "Date": "2021-01-05",
            },
        ]
)
result_df = clear_data(
    df.copy()
)  # 輸入函數, 得到結果
expected_df = pd.DataFrame(
    [
        {
            "StockID": "0050",
            "TradeVolume": "4962514",
            "Transaction": "6179",
            "TradeValue": "616480760",
            "Open": "124.20",
            "Max": "124.65",
            "Min": "123.75",
            "Close": "124.60",
            "Change": "0.25",
            "Date": "2021-01-05",
        },
        {
            "StockID": "0051",
            "TradeVolume": "175269",
```

```
                    "Transaction": "44",
                    "TradeValue": "7827387",
                    "Open": "44.60",
                    "Max": "44.74",
                    "Min": "44.39",
                    "Close": "44.64",
                    "Change": "0.04",
                    "Date": "2021-01-05",
            },
            {
                    "StockID": "0052",
                    "TradeVolume": "1536598",
                    "Transaction": "673",
                    "TradeValue": "172232526",
                    "Open": "112.10",
                    "Max": "112.90",
                    "Min": "111.15",
                    "Close": "112.90",
                    "Change": "0.8",
                    "Date": "2021-01-05",
            },
        ]
)
# 預期結果, 做完資料清理
# 將原先的會計數字, 如 1,536,598
# 轉換為一般數字 1536598
assert (
    pd.testing.assert_frame_equal(
        result_df, expected_df
    )
    is None
)  # 檢查, 執行結果 == 預期結果
```

```python
def test_colname_zh2en():
    #  準備好 input 的假資料
    result_df = pd.DataFrame(
        [
            {
                0: "0050",
                1: "元大台灣50",
                2: "4,962,514",
                3: "6,179",
                4: "616,480,760",
                5: "124.20",
                6: "124.65",
                7: "123.75",
                8: "124.60",
                9: "<p style= color:red>+</p>",
                10: "0.25",
                11: "124.55",
                12: "123",
                13: "124.60",
                14: "29",
                15: "0.00",
            },
            {
                0: "0051",
                1: "元大中型100",
                2: "175,269",
                3: "44",
                4: "7,827,387",
                5: "44.60",
                6: "44.74",
```

```
                    7: "44.39",
                    8: "44.64",
                    9: "<p style= color:red>+</p>",
                    10: "0.04",
                    11: "44.64",
                    12: "20",
                    13: "44.74",
                    14: "2",
                    15: "0.00",
                },
        ]
)
colname = [
        "證券代號",
        "證券名稱",
        "成交股數",
        "成交筆數",
        "成交金額",
        "開盤價",
        "最高價",
        "最低價",
        "收盤價",
        "漲跌(+/-)",
        "漲跌價差",
        "最後揭示買價",
        "最後揭示買量",
        "最後揭示賣價",
        "最後揭示賣量",
        "本益比",
]
result_df = colname_zh2en(
        result_df.copy(), colname
```

```
)  # 輸入函數, 得到結果
expected_df = pd.DataFrame(
    [
        {
            "StockID": "0050",
            "TradeVolume": "4,962,514",
            "Transaction": "6,179",
            "TradeValue": "616,480,760",
            "Open": "124.20",
            "Max": "124.65",
            "Min": "123.75",
            "Close": "124.60",
            "Dir": "<p style= color:red>+</p>",
            "Change": "0.25",
        },
        {
            "StockID": "0051",
            "TradeVolume": "175,269",
            "Transaction": "44",
            "TradeValue": "7,827,387",
            "Open": "44.60",
            "Max": "44.74",
            "Min": "44.39",
            "Close": "44.64",
            "Dir": "<p style= color:red>+</p>",
            "Change": "0.04",
        },
    ]
)
# 預期結果, 將 raw data , 包含中文欄位,
# 轉換成英文欄位, 以便存進資料庫
assert (
```

```
        pd.testing.assert_frame_equal(
            result_df, expected_df
        )
        is None
    )  # 檢查, 執行結果 == 預期結果


def test_twse_header():
    result = twse_header()
    expected = {
        "Accept": "application/json, text/javascript, */*; q=0.01",
        "Accept-Encoding": "gzip, deflate",
        "Accept-Language": "zh-TW,zh;q=0.9,en-US;q=0.8,en;q=0.7",
        "Connection": "keep-alive",
        "Host": "www.twse.com.tw",
        "Referer": "https://www.twse.com.tw/zh/page/trading/exchange/
MI_INDEX.html",
        "User-Agent": "Mozilla/5.0 (Windows NT 10.0; Win64;
x64) AppleWebKit/537.36 (KHTML, like Gecko) Chrome/71.0.3578.98
Safari/537.36",
        "X-Requested-With": "XMLHttpRequest",
    }
    assert result == expected


def test_tpex_header():
    result = tpex_header()
    expected = {
        "Accept": "application/json, text/javascript, */*; q=0.01",
        "Accept-Encoding": "gzip, deflate",
        "Accept-Language": "zh-TW,zh;q=0.9,en-US;q=0.8,en;q=0.7",
        "Connection": "keep-alive",
```

```
        "Host": "www.tpex.org.tw",
        "Referer": "https://www.tpex.org.tw/web/stock/aftertrading/
otc_quotes_no1430/stk_wn1430.php?l=zh-tw",
        "User-Agent": "Mozilla/5.0 (Windows NT 10.0; Win64; x64)
AppleWebKit/537.36 (KHTML, like Gecko) Chrome/73.0.3683.103
Safari/537.36",
        "X-Requested-With": "XMLHttpRequest",
    }
    assert result == expected


def test_set_column():
    # 準備好 input 的假資料
    df = pd.DataFrame(
        [
            {
                0: "00679B",
                2: "44.91",
                3: "-0.08",
                4: "45.00",
                5: "45.00",
                6: "44.85",
                7: "270,000",
                8: "12,127,770",
                9: "147",
            },
            {
                0: "00687B",
                2: "47.03",
                3: "-0.09",
                4: "47.13",
                5: "47.13",
```

```
                6: "47.00",
                7: "429,000",
                8: "20,181,570",
                9: "39",
            },
            {
                0: "00694B",
                2: "37.77",
                3: "-0.07",
                4: "37.84",
                5: "37.84",
                6: "37.72",
                7: "343,000",
                8: "12,943,630",
                9: "35",
            },
        ]
    )
result_df = set_column(
        df
    )  # 輸入函數, 得到結果
expected_df = pd.DataFrame(
        [
            {
                "StockID": "00679B",
                "Close": "44.91",
                "Change": "-0.08",
                "Open": "45.00",
                "Max": "45.00",
                "Min": "44.85",
                "TradeVolume": "270,000",
                "TradeValue": "12,127,770",
```

```
            "Transaction": "147",
        },
        {
            "StockID": "00687B",
            "Close": "47.03",
            "Change": "-0.09",
            "Open": "47.13",
            "Max": "47.13",
            "Min": "47.00",
            "TradeVolume": "429,000",
            "TradeValue": "20,181,570",
            "Transaction": "39",
        },
        {
            "StockID": "00694B",
            "Close": "37.77",
            "Change": "-0.07",
            "Open": "37.84",
            "Max": "37.84",
            "Min": "37.72",
            "TradeVolume": "343,000",
            "TradeValue": "12,943,630",
            "Transaction": "35",
        },
    ]
)
# 預期結果，根據資料的位置，設置對應的欄位名稱
assert (
    pd.testing.assert_frame_equal(
        result_df, expected_df
    )
    is None
```

```python
    ) # 檢查, 執行結果 == 預期結果


def test_crawler_twse_data9():
    """
    測試在證交所, 2021 正常爬到資料的情境,
    data 在 response 底下的 key, data9
    一般政府網站, 長時間的資料, 格式常常不一致
    """
    result_df = crawler_twse(
        date="2021-01-05"
    ) # 執行結果
    assert (
        len(result_df) == 20596
    ) # 檢查, 資料量是否正確
    assert list(result_df.columns) == [
        "StockID",
        "TradeVolume",
        "Transaction",
        "TradeValue",
        "Open",
        "Max",
        "Min",
        "Close",
        "Change",
        "Date",
    ] # 檢查, 資料欄位是否正確


def test_crawler_twse_data8():
    """
    測試在證交所, 2008 正常爬到資料的情境, 時間不同, 資料格式不同
```

```
    data 在 response 底下的 key, data8
    一般政府網站, 長時間的資料, 格式常常不一致
    """
    result_df = crawler_twse(
        date="2008-01-04"
    )
    assert (
        len(result_df) == 2760
    )  # 檢查, 資料量是否正確
    assert list(result_df.columns) == [
        "StockID",
        "TradeVolume",
        "Transaction",
        "TradeValue",
        "Open",
        "Max",
        "Min",
        "Close",
        "Change",
        "Date",
    ]  # 檢查, 資料欄位是否正確


def test_crawler_twse_no_data():
    """
    測試沒 data 的時間點, 爬蟲是否正常
    """
    result_df = crawler_twse(
        date="2000-01-04"
    )
    assert (
        len(result_df) == 0
```

```
)  # 沒 data, 回傳 0
# 沒 data, 一樣要回傳 pd.DataFrame 型態
assert isinstance(
    result_df, pd.DataFrame
)


def test_crawler_twse_error(mocker):
    """
    測試, 情境為, 對方網站回傳例外狀況, 或是被 ban IP 時, 爬蟲是否會失敗

    這邊使用特別的技巧, mocker,
    因為在測試階段, 無法保證對方 server 一定會給錯誤的結果
    因此使用 mocker, 對 requests 做"替換", 換成設定的結果
    如下
    """
    # 將特定路徑下的 requests 替換掉
    mock_requests = mocker.patch(
        "financialdata.crawler.taiwan_stock_price.requests"
    )
    # 將 requests.get 的回傳值 response, 替換掉成 ""
    # 如此一來, 當在測試爬蟲時,
    # 發送 requests 得到的 response, 就會是 ""
    mock_requests.get.return_value = ""
    result_df = crawler_twse(
        date="2000-01-04"
    )
    assert (
        len(result_df) == 0
    )  # 沒 data, 回傳 0
    # 沒 data, 一樣要回傳 pd.DataFrame 型態
    assert isinstance(
```

```python
        result_df, pd.DataFrame
    )


def test_crawler_tpex_success():
    """
    測試櫃買中心, 爬蟲成功時的狀況
    """
    result_df = crawler_tpex(
        date="2021-01-05"
    )  # 執行結果
    assert (
        len(result_df) == 6609
    )  # 檢查, 資料量是否正確
    assert list(result_df.columns) == [
        "StockID",
        "Close",
        "Change",
        "Open",
        "Max",
        "Min",
        "TradeVolume",
        "TradeValue",
        "Transaction",
        "Date",
    ]


def test_crawler_tpex_no_data():
    """
    測試沒 data 的時間點, 爬蟲是否正常
    """
```

```python
    result_df = crawler_tpex(
        date="2021-01-01"
    )
    assert (
        len(result_df) == 0
    )  # 沒 data, 回傳 0
    # 沒 data, 一樣要回傳 pd.DataFrame 型態
    assert isinstance(
        result_df, pd.DataFrame
    )


def test_convert_change():
    # 準備好 input 的假資料
    df = pd.DataFrame(
        [
            {
                "StockID": "0050",
                "TradeVolume": "4,680,733",
                "Transaction": "5,327",
                "TradeValue": "649,025,587",
                "Open": "139.00",
                "Max": "139.20",
                "Min": "138.05",
                "Close": "138.30",
                "Dir": "<p style= color:green>-</p>",
                "Change": "0.65",
                "Date": "2021-07-01",
            },
            {
                "StockID": "0051",
                "TradeVolume": "175,374",
```

```
                "Transaction": "120",
                "TradeValue": "10,152,802",
                "Open": "58.20",
                "Max": "59.10",
                "Min": "57.40",
                "Close": "57.90",
                "Dir": "<p style= color:green>-</p>",
                "Change": "0.30",
                "Date": "2021-07-01",
            },
            {
                "StockID": "0052",
                "TradeVolume": "514,042",
                "Transaction": "270",
                "TradeValue": "64,127,738",
                "Open": "125.00",
                "Max": "125.20",
                "Min": "124.35",
                "Close": "124.35",
                "Dir": "<p style= color:green>-</p>",
                "Change": "0.65",
                "Date": "2021-07-01",
            },
        ]
    )
    result_df = convert_change(
        df
    )  # 執行結果
    expected_df = pd.DataFrame(
        [
            {
                "StockID": "0050",
```

```
        "TradeVolume": "4,680,733",
        "Transaction": "5,327",
        "TradeValue": "649,025,587",
        "Open": "139.00",
        "Max": "139.20",
        "Min": "138.05",
        "Close": "138.30",
        "Change": -0.65,
        "Date": "2021-07-01",
    },
    {
        "StockID": "0051",
        "TradeVolume": "175,374",
        "Transaction": "120",
        "TradeValue": "10,152,802",
        "Open": "58.20",
        "Max": "59.10",
        "Min": "57.40",
        "Close": "57.90",
        "Change": -0.3,
        "Date": "2021-07-01",
    },
    {
        "StockID": "0052",
        "TradeVolume": "514,042",
        "Transaction": "270",
        "TradeValue": "64,127,738",
        "Open": "125.00",
        "Max": "125.20",
        "Min": "124.35",
        "Close": "124.35",
        "Change": -0.65,
```

```
                    "Date": "2021-07-01",
            },
        ]
    )
    # 預期結果,
    # 將 Dir (正負號) 與 Change (漲跌幅) 結合
    assert (
        pd.testing.assert_frame_equal(
            result_df, expected_df
        )
        is None
    )  # 檢查, 執行結果 == 預期結果



def test_convert_date():
    date = (
        "2021-07-01"  #  準備好 input 的假資料
    )
    result = convert_date(date)  # 執行結果
    expected = "110/07/01"  # 預期結果
    assert (
        result == expected
    )  # 檢查, 執行結果 == 預期結果



def test_crawler_twse():
    # 測試證交所爬蟲, end to end test
    result_df = crawler(
        parameter={
            "date": "2021-01-05",
            "data_source": "twse",
        }
```

```
    )
    result_df = check_schema(
        result_df, "TaiwanStockPrice"
    )
    assert len(result_df) > 0


def test_crawler_tpex():
    # 測試櫃買中心爬蟲, end to end test
    result_df = crawler(
        parameter={
            "date": "2021-01-05",
            "data_source": "tpex",
        }
    )
    result_df = check_schema(
        result_df, "TaiwanStockPrice"
    )
    assert len(result_df) > 0
```

這裡用到進階技巧，**mocker**，可參考上面 test_crawler_twse_error 函數，
該情境主要是因為在測試階段，需要發送 Requests，但無法保證對方回傳
的結果。

舉例來說，爬蟲的網站在半夜 12 點，會固定維護，當下發送 Requests
時，得到的結果與其他時間點不同，因此平常的時段，測不到這種情
境，那怎麼辦呢？

因應上面情境，使用 mocker 替換掉 requests，在測試階段，將 requests
回傳的結果寫定，固定成半夜 12 點回傳的結果，這樣可以保證在測試當
下，是模擬半夜 12 點的情境，並針對這個 Case，做例外處理。

當寫好測試後，可以觀察 8.1.2 中 VS Code 的插件功能，會顯示所有的測
試，並可針對特定測試去執行。

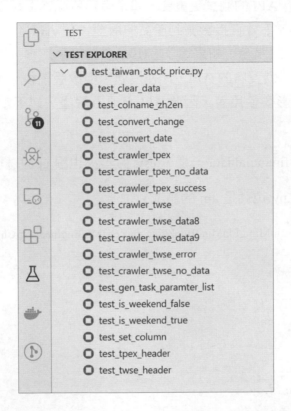

該爬蟲專案，只將 taiwan_stock_price.py 的測試覆蓋率做到 100%，其他
部分，就留給讀者做後續的練習，只要增加測試後，持續執行以下指令：

```
pipenv run pytest --cov-report term-missing --cov-config=.coveragerc
--cov=./financialdata/ tests/
```

就能看著測試覆蓋率慢慢提高，盡可能將覆蓋率到 80%、90% 以上，會
是比較好的習慣。下一章節，將介紹 API 的測試。

8.1.5 API 單元測試範例

相對爬蟲來說，API 的測試更重要，如果爬蟲新版本上線後壞了，頂多當下做修正即可，不會直接影響到產品的服務。但 API 不同，因為 API 是直接面對 User 的服務，如果壞了，User 會直接受影響，不論是 Backend 的 API，或是提供資料的 API，即使沒壞，但改版前後，結構 Schema 不同，也會大大影響使用者，因此 API 必須更嚴謹，以下拿 7.6 的 API 來做測試範例。

架構如下，跟 financialdata 一樣，tests 資料夾也跟 api 在同一層。

tests/test_main.py 也就是 api 的測試。

https://github.com/FinMind/FinMindBook/tree/master/DataEngineering/Chapter8/8.1.5

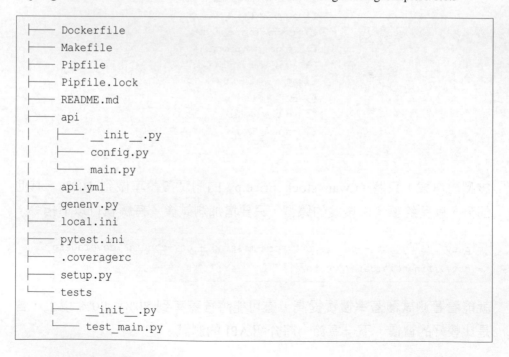

```
├──── Dockerfile
├──── Makefile
├──── Pipfile
├──── Pipfile.lock
├──── README.md
├──── api
│    ├──── __init__.py
│    ├──── config.py
│    └──── main.py
├──── api.yml
├──── genenv.py
├──── local.ini
├──── pytest.ini
├──── .coveragerc
├──── setup.py
└──── tests
     ├──── __init__.py
     └──── test_main.py
```

由於 API 比較單純，因此測試也不會太複雜，是個方便作為入門的範例。程式如下：

同樣，先安裝測試 Package：

```
pipenv install pytest-cov==2.11.1 pytest-mock==3.5.1
```

用以下指令進行測試與覆蓋率計算：

```
pipenv run pytest --cov-report term-missing --cov-config=.coveragerc
--cov=./api/ tests/
```

測試結果如下圖，這裡很容易就達到 100% 覆蓋率，因為相對爬蟲來說，API 只是從資料庫撈資料出來而已，比較少做邏輯處理，因此測試很容易撰寫：

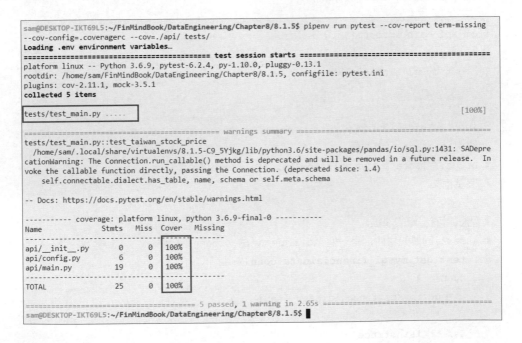

```
sam@DESKTOP-IKT69L5:~/FinMindBook/DataEngineering/Chapter8/8.1.5$ pipenv run pytest --cov-report term-missing
--cov-config=.coveragerc --cov=./api/ tests/
Loading .env environment variables…
=================================== test session starts ===================================
platform linux -- Python 3.6.9, pytest-6.2.4, py-1.10.0, pluggy-0.13.1
rootdir: /home/sam/FinMindBook/DataEngineering/Chapter8/8.1.5, configfile: pytest.ini
plugins: cov-2.11.1, mock-3.5.1
collected 5 items

tests/test_main.py .....                                                            [100%]

==================================== warnings summary =====================================
tests/test_main.py::test_taiwan_stock_price
  /home/sam/.local/share/virtualenvs/8.1.5-C9_5Yjkg/lib/python3.6/site-packages/pandas/io/sql.py:1431: SADepre
cationWarning: The Connection.run_callable() method is deprecated and will be removed in a future release.  In
voke the callable function directly, passing the Connection. (deprecated since: 1.4)
    self.connectable.dialect.has_table, name, schema or self.meta.schema

-- Docs: https://docs.pytest.org/en/stable/warnings.html

---------- coverage: platform linux, python 3.6.9-final-0 -----------
Name               Stmts   Miss  Cover   Missing
--------------------------------------------------
api/__init__.py        0      0   100%
api/config.py          6      0   100%
api/main.py           19      0   100%
--------------------------------------------------
TOTAL                 25      0   100%
============================ 5 passed, 1 warning in 2.65s =================================
sam@DESKTOP-IKT69L5:~/FinMindBook/DataEngineering/Chapter8/8.1.5$ █
```

以下將展示測試範例：

https://github.com/FinMind/FinMindBook/blob/master/DataEngineering/
Chapter8/8.1.5/tests/test_main.py

test_main.py，關於每個測試的情境、細節，本書都寫在註解上：

```python
import time
from multiprocessing import Process

import pytest
import requests
import uvicorn
from fastapi.testclient import (
    TestClient,
)
from sqlalchemy import engine

from api.main import (
    app,
    get_mysql_financialdata_conn,
)


client = TestClient(app)
# 使用 fastapi 官方教學
# https://fastapi.tiangolo.com/tutorial/testing/
# 測試框架

# 測試對資料庫的連線,
# assert 回傳的物件, 是一個 sqlalchemy 的 connect 物件
def test_get_mysql_financialdata_conn():
    conn = (
        get_mysql_financialdata_conn()
    )
    assert isinstance(
```

```
        conn, engine.Connection
    )


# 測試對 'http://127.0.0.1:5000/' 頁面發送 request,
# 得到的回應 response 的狀態 status_code, json data
def test_read_root():
    response = client.get("/")
    assert response.status_code == 200
    assert response.json() == {
        "Hello": "World"
    }


# 測試對 'http://127.0.0.1:5000/taiwan_stock_price' 頁面發送 request,
# 並帶 stock_id, start_date, end_date 參數
# 得到的回應 response 的狀態 status_code, json data
def test_taiwan_stock_price():
    response = client.get(
        "/taiwan_stock_price",
        params=dict(
            stock_id="2330",
            start_date="2021-04-01",
            end_date="2021-04-01",
        ),
    )
    assert response.status_code == 200
    # 以下特別重要, 需要把 response 結果寫死
    # 避免未來 api schema 改變時, 影響 api 的使用者
    # 例如 Open 目前 response 是 float, 598.0,
    # 未來就不能改成 int, 598, 這是不被允許的
    # 因為 float -> int, 會大大影響使用者進行資料處理
    # 同理, int -> float 也是禁止的,
    # 資料型態在一開始就必須決定好
```

```python
    assert response.json() == {
        "data": [
            {
                "StockID": "2330",
                "TradeVolume": 45972766,
                "Transaction": 48170,
                "TradeValue": 27520742963,
                "Open": 598.0,
                "Max": 602.0,
                "Min": 594.0,
                "Close": 602.0,
                "Change": 15.0,
                "Date": "2021-04-01",
            }
        ]
    }


# end to end 測試, 模擬真實使用 request 套件發送請求
# 使用 Process 開另一個進程, 模擬啟動 api
# 之後會在主進程, 對此 api 發送 request
@pytest.fixture(scope="module")
def setUp():
    proc = Process(
        target=uvicorn.run,
        args=(app,),
        kwargs={
            "host": "127.0.0.1",
            "port": 5000,
            "log_level": "info",
        },
        daemon=True,
```

```
    )
    proc.start()
    time.sleep(1)
    return 1



# 測試對 api 發送 requests,
# assert 回傳結果是 {"Hello": "World"}
def test_index(setUp):
    response = requests.get(
        "http://127.0.0.1:5000"
    )
    assert response.json() == {
        "Hello": "World"
    }
# 測試對 api 發送 requests,
# assert 回傳的 data, 這裡把真實的 case 寫下來
# 跟 test_taiwan_stock_price 概念一樣,
# 唯一的差別是, 這裡是真實場景, 對 api 發送 requests
def test_TaiwanStockPriceID(setUp):
    payload = {
        "stock_id": "2330",
        "start_date": "2021-04-01",
        "end_date": "2021-04-01",
    }
    res = requests.get(
        "http://127.0.0.1:5000/taiwan_stock_price",
        params=payload,
    )
    resp = res.json()["data"]
    assert resp == [
        {
```

```
        "StockID": "2330",
        "TradeVolume": 45972766,
        "Transaction": 48170,
        "TradeValue": 27520742963,
        "Open": 598.0,
        "Max": 602.0,
        "Min": 594.0,
        "Close": 602.0,
        "Change": 15.0,
        "Date": "2021-04-01",
    }
]
```

這裡除了使用 FastAPI 推薦的測試框架外，也同時測試，真實架設 API Server，並發送 Request 的情境，因此使用 Process 額外開一個進程，讓讀者能夠在同一個測試中，同時架 Server 跟發送 Request。

跟 8.1.4 一樣，也可以用 VS Code 提供的測試插鍵來做單一測試。

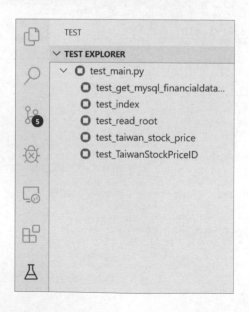

以上測試都是手動進行，但人都會健忘，可能在某次新版本上，忘記跑測試，那怎麼辦呢？或是另一個情境，在多人開發上，可能某些人沒跑測試就 Merge，這也是一個問題，這是工作上真實的案例，如果將以上測試，加入 GIT 中，在每次發送 Pull Request or Merge Request 時，自動跑測試，這是不是使整個開發流程更順暢呢？

下個章節，將介紹 Continuous Integration（CI），持續性整合。

CICD 持續性整合、部屬

什麼是 CICD？

CICD 原文是 Continuous Integration、Continuous Delivery，簡單來說，
就是**自動化程式開發流程**。持續將新程式、功能，**整合進原先的專案中**，
這就是 CI，而 CD 就是持續將 CI 整合過後的新版本，部屬到正式（產
品）環境上。

實務上的情境，團隊成員開發新功能後，經過 CICD，自動測試並部屬新
版本到正式環境中，讓新功能交付這件事，有個 SOP 流程，且自動執行。

9.2　CI 持續性整合

可能有些人會認為，CI 就是單純跑測試，實際上 CI 是，持續性整合，在多人開發上，持續將團隊開發的程式、新功能，整合到主產品上，且將流程自動化。新的功能越快整合到產品上，整個開發流程會更加快速。

某些工程師可能會提出疑問，CI 不是 DevOps 的工作嗎？為什麼一般工程師需要了解？我只要專心開發新功能、新的分析結果就好了。筆者認為，工程師必須要有專業，要對自己開發的程式負責，有獨立作業的能力，不能什麼都依靠別人，而最好的負責方式就是，確保自己的程式不要發生 Error，特別是做產品、服務，來自 User 的壓力是很可怕的。

本書的主軸是，以 Data 為起點做一個產品。試想一下，如果離開公司，沒有 QA、沒有 DevOps、沒有 SRE，你私底下開發 Side Project、產品，有辦法保證穩定性嗎？

CI 可以協助軟體開發的流程，而 CI 工具有很多種，以下將介紹 Gitlab-CI，並以 8.1.4 的爬蟲專案為例。

9.3　Gitlab-CI、以爬蟲專案為例

這裡使用 Gitlab-CI，無法在 GitHub 上進行，必須使用 Gitlab，因此專案改到以下連結進行。

https://gitlab.com/FinMindBook/financialdata

首先，需要 **.gitlab-ci.yml** 設定檔，如下，設定檔每個步驟，都寫在註解中，這裡要注意一點，CI 測試會在 Docker 中運行，確保隔絕一切外在因素，讓測試與實際運作在同一個 Docker 環境底下：

https://gitlab.com/FinMindBook/financialdata/-/blob/feature/9.3/.gitlab-ci.yml

```
.gitlab-ci.yml

stages:
  # CI pipeline
  # 一般開發流程
  # test 測試過了 ->
  #   merge 進 stage/master ->
  #   build 建立 docker image ->
  #   deploy 部屬服務
  # 會根據此順序依序進行
  # test -> build -> deploy
  - test
  - build
  - deploy


# CI 名稱，test 步驟
test-crawler:
    # 在 stage 設定 CI pipeline 順序
    # 這裡 pipeline 步驟是 test
    stage: test
    # 運行測試的 docker image
    # 由於我們 Dockerfile 基於
    # ubuntu:18.04 下去做設定
    # 因此這裡測試也使用同樣的 image

    # 本書前一版使用 continuumio/miniconda3:4.3.27
    # 但此 image 中的 Linux 系統 - Debian 版本太舊
```

```
# 因此改用 ubuntu 系統
image: ubuntu:18.04
before_script:
  # 測試事前準備的指令
  # 需要先安裝環境
  - apt-get update && apt-get install python3.6 -y && apt-get install
python3-pip -y
  - pip3 install pipenv==2020.6.2
  - export LC_ALL=C.UTF-8 LANG=C.UTF-8
  - pipenv sync
  - python3 genenv.py
script:
  # 實際測試指令
  - pipenv run pytest --cov-report term-missing --cov-config=
.coveragerc --cov=./financialdata/ tests/
only:
    # 設定只有在 merge requests 的情況下
    # 才會運作此 CI
    refs:
      - merge_requests
```

- stage：設定 CICD pipeline 步驟，test。
- image：設定與 Dockerfile 相同的 image。
- before_script：事前指令，這邊進行環境安裝。
- script：主要指令，在這進行測試。
- only：設定在什麼情況下，才會執行該 pipeline，這邊設定在 merge_
  request 的情況下跑測試。

以上展示 CI Test 測試步驟，而整個 CICD 流程還包含，Build，建立
Docker Image，Deploy，部屬新版本，這在後續章節會一一介紹。

那 Test 實際上怎麼運作呢？當開發一個新功能或是對於程式做改動，對於原本的程式，都有風險，可能會導致原本的程式出錯，而測試就是避免這個問題，因此設定在 Merge 進 master ／ stage 前（refs: merge_request），會自動進行測試，不用再手動。

實際情境如下圖：

https://gitlab.com/FinMindBook/financialdata/-/merge_requests/16

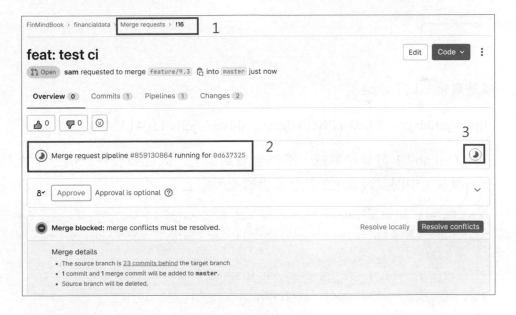

上圖 1，開發完新 Feature 後，建立 Merge Requests，在圖 2 的地方，就會自動進行 CI 測試，而測試執行的步驟，就是根據 .gitlab-ci.yml 的設定（script）去運行，可以點擊圖 3，進去觀察運作是否正常（點擊圖 3 後，會跳轉到以下頁面，需再點擊 test-crawler）。

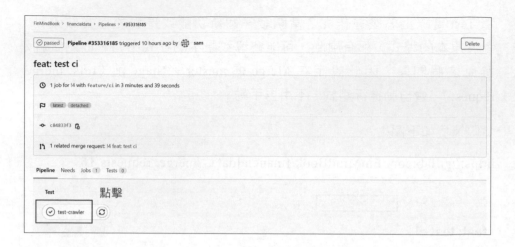

或是直接點以下連結：

https://gitlab.com/FinMindBook/financialdata/-/jobs/1504132413

由於 GitLab-CI 背景是黑底，不適合放到書上，因此複製運行指令，並一一講解（中間安裝環境的訊息將會省略不貼上）。

```
Running with gitlab-runner 15.9.0~beta.115.g598a7c91 (598a7c91)
  on blue-5.shared.runners-manager.gitlab.com/default -AzERasQ, system
ID: s_8a38c517a741
  feature flags: FF_USE_IMPROVED_URL_MASKING:true
Preparing the "docker+machine" executor
Using Docker executor with image ubuntu:18.04 ...
Pulling docker image ubuntu:18.04 ...
```

GitLab-CI 是在 Docker 環境中進行，需要 pull ubuntu:18.04，這 Image 是在 .gitlab-ci.yml 中設定的。

```
$ apt-get update && apt-get install python3.6 -y && apt-get install
python3-pip -y
```

在 .gitlab-ci.yml 中，before_script，寫明需要先安裝環境，因此 CI 正在
按照設定的指令運作。

```
$ pip3 install pipenv==2020.6.2
```

before_script 第二個步驟，使用安裝 pipenv。

```
$ export LC_ALL=C.UTF-8 LANG=C.UTF-8
```

before_script 的第三個步驟，設定環境變數。

```
$ pipenv sync
```

before_script 的第四個步驟，安裝 Package。

```
$ python3 genenv.py
```

before_script 的第五個步驟，使用 genenv.py 建立環境變數 .env。

```
$ pipenv run pytest --cov-report term-missing --cov-config=.coveragerc
--cov=./financialdata/ tests/
```

此步驟，是 .gitlab-ci.yml 中 script 的指令，也就是前面章節使用的測試
指令。最後運作結果如下：

```
$ pipenv run pytest --cov-report term-missing --cov-config=.coveragerc
--cov=./financialdata/ tests/
Loading .env environment variables...
=========================== test session starts ============================
platform linux -- Python 3.6.2, pytest-6.2.4, py-1.10.0, pluggy-0.13.1
rootdir: /builds/FinMindBook/financialdata, configfile: pytest.ini
plugins: cov-2.11.1, mock-3.5.1
collected 18 items
tests/crawler/test_taiwan_stock_price.py ...................
[100%]
```

```
----------- coverage: platform linux, python 3.6.2-final-0 -----------
Name                                        Stmts  Miss  Cover Missing
----------------------------------------------------------------------
financialdata/__init__.py                       0     0   100%
financialdata/backend/__init__.py               0     0   100%
financialdata/backend/db/__init__.py            5     5     0%  1-8
financialdata/backend/db/clients.py             7     7     0%  1-18
financialdata/backend/db/db.py                  9     9     0%  1-19
financialdata/backend/db/router.py             30    30     0%  1-55
financialdata/config.py                        10    10     0%  1-13
financialdata/crawler/__init__.py               0     0   100%
financialdata/crawler/taiwan_futures_daily.py  51    51     0%  1-141
financialdata/crawler/taiwan_stock_price.py    96     0   100%
financialdata/producer.py                      15    15     0%  1-30
financialdata/scheduler.py                     17    17     0%  1-34
financialdata/schema/__init__.py                0     0   100%
financialdata/schema/dataset.py                34     0   100%
financialdata/tasks/__init__.py                 0     0   100%
financialdata/tasks/task.py                     8     8     0%  1-19
financialdata/tasks/worker.py                   4     4     0%  1-13
----------------------------------------------------------------------
TOTAL                                         286   156    45%
==================== 18 passed in 71.82s (0:01:11) ====================
```

跟 8.1.4 的結果一致，唯一的差別是，這是在 GitLab-CI 自動運作，不需要每次都手動。流程自動化，就是本書的目標，這可以加快開發速度與穩定性。

接下來，將介紹 Build 的步驟，建立 Docker Image，並 Push 推到 Docker Hub。

9.4　Gitlab-CI，建立 Docker Image

9.4.1 Gitlab-Runner

根據 Gitlab 官方文件：

https://docs.gitlab.com/12.10/ee/ci/docker/using_docker_build.html

在 CICD 階段，要建立 Docker Image 時，需要先設定 GitLab-Runner，那 GitLab-Runner 是什麼呢？

簡單來說，是幫你運作 CI 指令的機器，工人（Worker）的概念，以 9.3 的例子來說，Merge Request 會觸發 CI，這時 GitLab 就會發送任務給 GitLab Runner，讓 Runner 去進行測試的 Job，這跟 5.5 介紹的 Celery，是很類似的概念，都是發送 Job 給 Worker 去執行，以下將講解如何設定 GitLab Runner。

以 financialdata 的 GitLab 專案為例：

https://gitlab.com/FinMindBook/financialdata

到 Settings 點選 CI/CD：

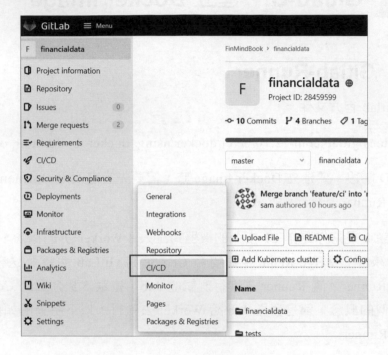

再到 Runners 的地方點選 Expand：

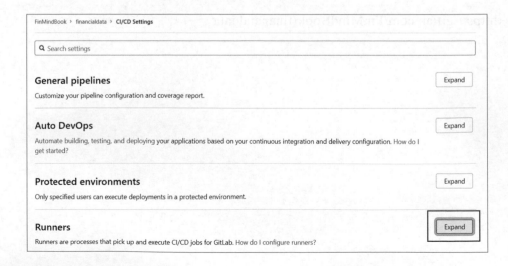

就會看到 runners 的設定，如下：

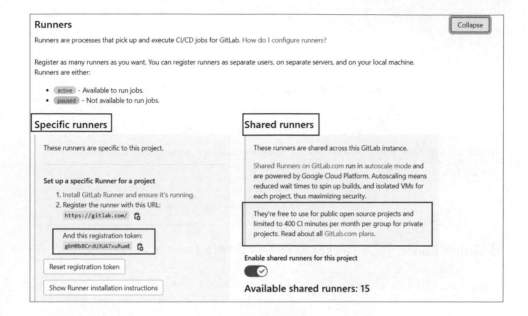

這裡有兩種 runner，**Specific runners** 和 **Shared runners**。

Specific runners 是讓使用者自行設定在特定機器上跑的 Runner，GitLab
官方也有提供教學。

Shared runners 是 GitLab 官方提供的免費 Runner，私人專案每個月有
400 分鐘的運行時間。

根據官方教學：

https://docs.gitlab.com/12.10/ee/ci/docker/using_docker_build.html#use-shell-executor

要使用 **Specific runners**，並設定 shell runner 在指定的機器上，首先，需
要安裝 Gitlab-Runner，指令如下。

```
sudo curl -L --output /usr/local/bin/gitlab-runner https://gitlab-runner-
downloads.s3.amazonaws.com/latest/binaries/gitlab-runner-linux-amd64

sudo chmod +x /usr/local/bin/gitlab-runner

sudo useradd --comment 'GitLab Runner' --create-home gitlab-runner
--shell /bin/bash

sudo gitlab-runner install --user=gitlab-runner --working-directory=/
home/gitlab-runner

sudo gitlab-runner start
```

將 Gitlab-Runner 加入 Docker 權限，以便操作 Docker：

```
sudo usermod -aG docker gitlab-runner
```

安裝完後，執行以下指令，註冊 Gitlab-Runner。

```
sudo gitlab-runner register --non-interactive --url
"https://gitlab.com/" --registration-token "gbH8b8CrdU3UA7xuRumE"
--executor "shell" --description "build_image" --tag-list "build_image"
```

執行畫面如下：

```
sam@localhost:~$ sudo gitlab-runner register --non-interactive --url "https://gitlab.com/" --registration-token "gbH8b8CrdU3UA7xuRumE" --executor "shell" --descri
ption "build_image" --tag-list "build_image"
Running in system-mode.

Registering runner... succeeded                 runner=gbH8b8Cr
Runner registered successfully. Feel free to start it, but if it's running already the config should be automatically reloaded!
sam@localhost:~$
```

最後顯示 succeeded，代表成功註冊，這時重新整理畫面：

https://gitlab.com/FinMindBook/financialdata/-/settings/ci_cd

就會看到，Specific runners 底下，多了一個 runner，這就是以上指令設
定的 runner。

這時 runner 會先顯示驚嘆號，代表正在建立 runner，並與 GitLab 做連
線，等個約 3~5 分鐘，就會成功了，如下：

註冊指令蠻長的,本書一一講解。

- register:顧名思義,就是註冊 GitLab-Runner。
- --non-interactive:非互動模式,互動模式指的是,在註冊 GitLab-Runner 的階段,會被詢問關於設定細節等多個問題,如果加入此指定,就可以像一鍵執行,中間不會再詢問(不會再互動)多個問題。
- --url:顧名思義,就是註冊的 url,這裡選 GitLab 的官方網址。
- --registration-token:**Specific runners** 底下提供的 token。
- --executor:指定在什麼環境下執行,這裡選擇 shell,也就是一般 Linux 原生環境下,根據官方文件,https://docs.gitlab.com/runner/executors/,還有 Docker、Kubemetes 等選擇,其中 Docker 在接下來的章節會用到。

- --description：關於此 Runner 的描述，像書中設定 build_image，表示這 Runner 是給建立 Image 使用的。
- --tag-list：關於此 Runner 的 Tag，一般情境，會進行 Unit Test、Build Image、Deploy 等等的 CICD，因此會建立多個 Runner，那在 .gitlab-ci.yml 中，要如何選擇特定的 Runner 執行 CICD？這時就是依靠 Tag。

建立好 Runner 後，下個章節，將講解如何使用 Runner，並在 CICD 階段，建立 Docker Image。

9.4.2 CICD 建立 Docker Image

在 7.5，書中介紹了 Docker Swarm，在此架構下，爬蟲、API 都是用 Docker 包裝好，部屬上線，如果要發布新版本，都是藉由 Docker 進行。所以在第二步驟，需要 Build Docker Image，才能發布新版本，以下將介紹如何在 CICD 底下，將 Build Image 這件事，達到自動化。

與 9.3 相同，在 .gitlab-ci.yml 寫入，建立 Docker Image 的設定，如下：

https://gitlab.com/FinMindBook/financialdata/-/blob/9.4.2/.gitlab-ci.yml

```
.gitlab-ci.yml
stages:
  # CI pipeline
  # 一般開發流程
  # test 測試過了 ->
  #   merge 進 stage/master ->
  #   build 建立 docker image ->
  #   deploy 部屬服務
  # 會根據此順序依序進行
  # test -> build -> deploy
  - test
```

```
  - build
  - deploy


# test-crawler 的部分略過


# CI 名稱, 建立 docker image
build-docker-image:
    # 在 stage 設定 CI pipeline 順序
    # 這裡 pipeline 步驟是 build
    stage: build
    before_script:
        # 由於 build 好 image 後，需要 push 到 docker hub
        # 需要事先登入
        # 登入部分，這裡使用 token 做登入，可參考以下連結
        # https://docs.docker.com/docker-hub/access-tokens/#create-an-
access-token
        # token 你可以使用明碼，或是存入 gitlab CICD Variables，用變數的方式引用
        # 如下
        # docker login -u "linsamtw" -p 7777-7777-7777-7777
        # 但實務上，不希望 token 公開，因此會採用以下方式
        - docker login -u "linsamtw" -p ${DOCKER_HUB_TOKEN}
    script:
        # 建立 image
        - docker build -f Dockerfile -t linsamtw/crawler:9.4 .
        # push
        - docker push linsamtw/crawler:9.4
    tags:
        # 設定使用前面建立的 runner 執行
        - build_image
    only:
        # 設定只有在 staging, master 階段
        # 才會 build image
```

```
refs:
  - staging
  - master
```

- stage：設定 CICD pipeline 步驟，build。
- image：不設定，因為這裡要使用 9.4.1 註冊的 GitLab-Runner。
- before_script：事前指令，因為需要 Push Docker Image，先做登入動作。
- script：主要指令，在這建立 Image 與 Push 到 Docker Hub。
- tag：設定要用哪個 GitLab-Runner 執行該段 Pipeline。
- only：設定在 staging、master 階段，執行 build pipeline。

這裡有一點需特別注意，由於需要 Push Docker Image 到 Docker Hub，需事先登入，這裡使用 Token 做登入，如下：

```
docker login -u "linsamtw" -p ${DOCKER_HUB_TOKEN}
```

實務上，為了安全性，關於密碼、金鑰、Token 等資訊，都會存到 CICD Variables 中，以下先介紹如何拿到 Docker Token，接著再講解如何在 GitLab 設定 CICD Variables。

Docker Token 申請方式，以下是官方教學：
https://docs.docker.com/docker-hub/access-tokens/#create-an-access-token

簡單來說，到以下連結：
https://hub.docker.com/settings/security

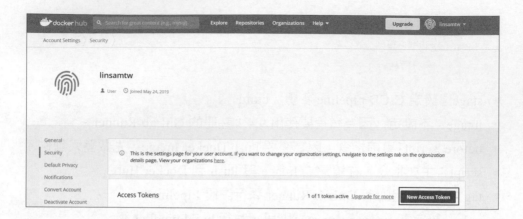

點選 New Access Token，就會看到以下畫面，接著建立你自己 Token。

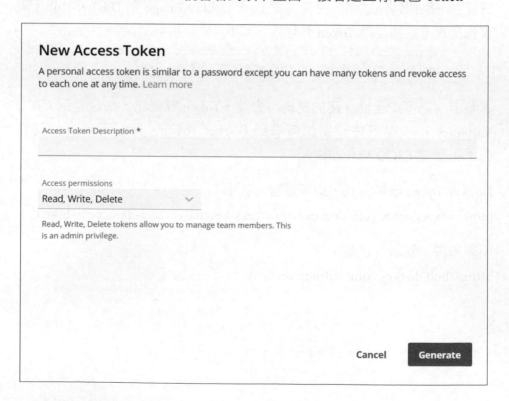

接著將 Token 存起來，到以下畫面，設定 Variables：

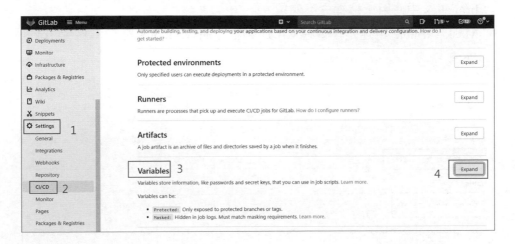

點選 Add variable：

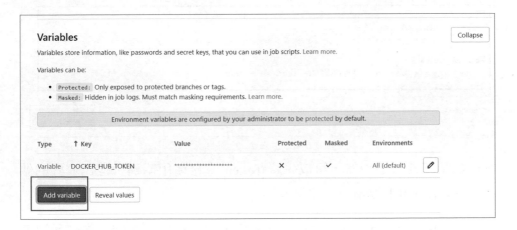

接著到 Key 的地方，輸入你想要的 Variable 名稱，這裡輸入 DOCKER_
HUB_TOKEN，接著在 Value 的地方，輸入你的 Token。

在 Flags 地方，有兩個選項 Protect variable、Mask variable。

- Protect variable 用途是，限定此 CICD Variables 只能在受保護的 branches 中使用，這裡先不勾選。
- Mask variable 用途是，會在 CICD 過程，將 Variables 遮住，避免 Token 外洩。

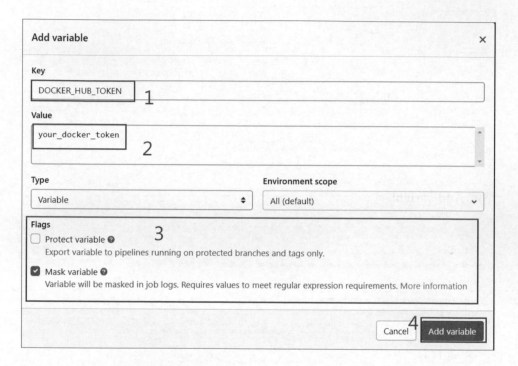

當然如果不想這麼複雜，可以先將 Token 直接使用明碼，但比較好的做法，還是用 CICD Variable 做管理。

接著當 Merge Request 合併到 master 後，讀者可能會想問，要去哪裡查看 Build Image 步驟？成功或是失敗的 Log ？

如下圖，先到 CICD，選擇 Jobs，可以看到 Name 為 build-docker-image，這裡的 Name，就是 .gitlab-ci.yml 中設定的名稱，接著點選圖 4，passed，就可以看到實際上，Build 執行的指令。

確認成功後，到 Docker Hub，就可以看到，CICD 上傳的 Docker Image。

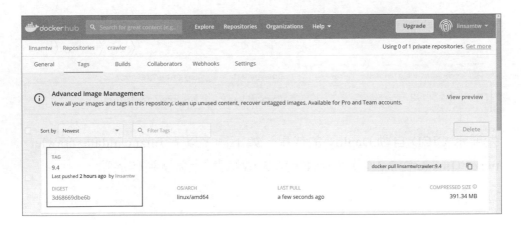

這時完成了 test、build 兩個步驟，且都達到自動化，不需要再手動跑測試、手動建立 Docker Image，下個章節，將展現第三步驟，部屬新版本，將 deploy 這步驟，也進行自動化。

9.5　Gitlab-CI，部屬新版本

首先，回到 7.5，本書建立了 Docker Swarm、並用 Portainer 作為 UI 介面，http://139.162.104.54:9000/#/services。

同樣以 financialdata 為例，financialdata_cralwer_twse、financialdata_scheduler，目前 Image 版本停留在，linsamtw/crawler:7.2.1，本章節將使用 CICD，協助 Deploy 部署到 Swarm 中，達到自動化。

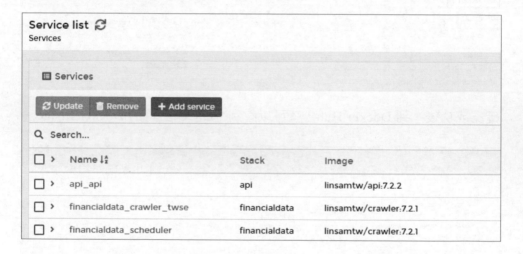

先展示 CICD 自動 Deploy 的成果，如下圖，以下 financialdata_crawler_twse，Image 從 7.2.1 更新版本到 9.5，那實際上怎麼運作的呢？

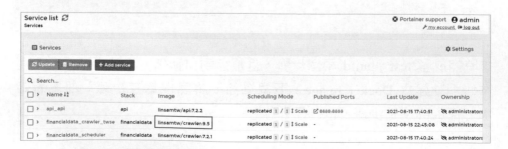

同樣，在 .gitlab.ci.yml 中，加入第三步驟，Deploy。

https://gitlab.com/FinMindBook/financialdata/-/blob/9.5/.gitlab-ci.yml

```
.gitlab-ci.yml

stages:
  # CI pipeline
  # 一般開發流程
  # test 測試過了 ->
  #   merge 進 stage/master ->
  #   build 建立 docker image ->
  #   deploy 部屬服務
  # 會根據此順序依序進行
  # test -> build -> deploy
  - test
  - build
  - deploy


# 一樣略過 test-crawler


# CI 名稱, 建立 docker image
build-docker-image:
    # 在 stage 設定 CI pipeline 順序
    # 這裡 pipeline 步驟是 build
    stage: build
    before_script:
        # 由於 build 好 image 後，需要 push 到 docker hub
        # 需要事先登入
        # 登入部分，這裡使用 token 做登入，可參考以下連結
        # https://docs.docker.com/docker-hub/access-tokens/#create-an-
access-token
        # token你可以使用明碼，或是存入gitlab CICD Variables，用變數的方式引用
```

```
    # 如下
    # docker login -u "linsamtw" -p 7777-7777-7777-7777
    # 但實務上，不希望 token 公開，因此會採用以下方式
    - docker login -u "linsamtw" -p ${DOCKER_HUB_TOKEN}
  script:
    # 建立 image
    - make build-image
    # push
    - make push-image
  tags:
    # 設定使用前面建立的 runner 執行
    - build_image
  only:
    # 設定只有在下 tag 後
    # 才會 build image
    - tags


# CI 名稱, deploy crawler
deploy-crawler:
  # 在 stage 設定 CI pipeline 順序
  # 這裡 pipeline 步驟是 deply
  stage: deploy
  script:
    # 部屬爬蟲
    - make deploy-crawler
  tags:
    # 設定使用前面建立的 runner 執行
    - build_image
  only:
    # 設定只有在下 tag 後
    # 才會 deploy
    - tags
```

本章節做了兩個更動：

1. only 部分，從 stage ／ master 改成 tags，跟發布／部屬新版本有關，在這種情境下，Merge 進 master 不會部屬，需要下 Tag 版本號，才會開始建立 Image，並部屬上線。主要是情境不同。
2. script 部分，改用 make 指令，主要是因為，每次 build image 的版本號會不同，在 9.4.2 章節，是直接寫死版本號，但在開發過程中，會持續建立新版本的 Docker Image，因此改用 Make，並在 Makefile 加入，版號更動這件事，如下。

這裡只列出本章節使用到的指令，其餘可直接參考連結：

https://gitlab.com/FinMindBook/financialdata/-/blob/9.5/Makefile

```Makefile
GIT_TAG := $(shell git describe --abbrev=0 --tags)


# 建立 docker image
build-image:
    docker build -f Dockerfile -t linsamtw/crawler:${GIT_TAG} .


# 推送 image
push-image:
    docker push linsamtw/crawler:${GIT_TAG}


# 部屬爬蟲
deploy-crawler:
    GIT_TAG=${GIT_TAG} docker stack deploy --with-registry-auth -c
crawler.yml financialdata


# 部屬 scheduler
deploy-scheduler:
```

```
    GIT_TAG=${GIT_TAG} docker stack deploy --with-registry-auth -c
scheduler.yml financialdata
```

- GIT_TAG：在這拿到 GitLab 的 Tag 編號，作為後續 Docker Image 的版本號。
- build-image：加入 ${GIT_TAG}，這樣才能引用 GIT_TAG 變數，根據此建立不同版號的 Image。
- push-image：一樣加入 ${GIT_TAG}，根據 GIT_TAG 推送不同版號的 Image。
- deploy-crawler： 加 入 GIT_TAB=${GIT_TAG}， 將 GIT_TAG 傳 入 crawler.yml。
- deploy-scheduler：同上。

crawler.yml 在 Image 的部分，同樣也做了更動，加入 GIT_TAG，作為版號控制，並在 Makefile 的指令中，將 GIT_TAG 傳進去，如下。

https://gitlab.com/FinMindBook/financialdata/-/blob/9.5/crawler.yml

crawler.yml

```
version: '3.0'
services:
  crawler_twse:
    image: linsamtw/crawler:${GIT_TAG}
    hostname: "twse"
    command: pipenv run celery -A financialdata.tasks.worker worker
--loglevel=info --concurrency=1  --hostname=%h -Q twse
    restart: always
    # swarm 設定
    deploy:
      mode: replicated
      replicas: 1
```

```
    placement:
      constraints: [node.labels.crawler_twse == true]
    environment:
      - TZ=Asia/Taipei
    networks:
      - my_network


networks:
  my_network:
    # 加入已經存在的網路
    external: true
```

完成以上更動並 Merge 到 master 後,新增一個 Tag,CICD 就會自動執行 build 與 deploy 步驟,如下:

https://gitlab.com/FinMindBook/financialdata/-/tags

新增好 Tag 後,再到 CI/CD 中的 Piplines,就能看到執行的步驟了,第一個是 build。

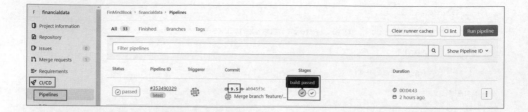

第二個是 deploy。

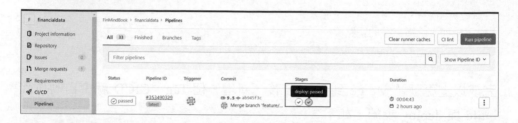

這時切換到 Portainer 頁面，就能看到，financialdata_crawler_twse 已經換成新的 Image 了。

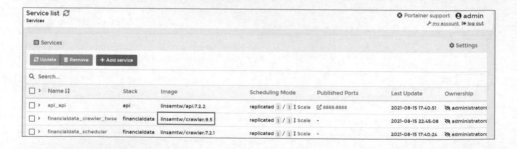

financialdata_scheduler 部分，讀者可自行仿造，或是直接參考以下連結：

https://gitlab.com/FinMindBook/financialdata/-/blob/9.5.2/.gitlab-ci.yml

成功更新畫面如下：

Q Search...					
□ › Name ↓↑	Stack	Image	Scheduling Mode	Published Ports
□ › api_api	api	linsamtw/api:7.2.2	replicated 1 / 1 ⇕ Scale	☑ 8888:8888
□ › financialdata_crawler_twse	financialdata	linsamtw/crawler:9.5.2	replicated 1 / 1 ⇕ Scale	-
□ › financialdata_scheduler	financialdata	linsamtw/crawler:9.5.2	replicated 1 / 1 ⇕ Scale	-

到此步驟，將 test - 測試、build - 建立 Image、deploy - 部屬上線，都完成自動化了，即使多人開發，也只需要按個按鈕，建立新的 Tag，其他成員也能自行換版，且保證通過測試，完成了 CICD 自動化流程。

下一章節，9.6，本書繼續介紹 API 專案的 CICD 流程。

9.6 Gitlab-CI、以 API 專案為例

9.6.1 Test - 測試

API 測試，包含對資料庫連線的 End to End 測試，但資料庫，一般都會設防火牆，只限特定 IP 做連線，必須使用 Gitlab-Runner 來執行測試，無法使用免費的 Share Runner。

上一章節，註冊了 shell 類型的 Gitlab Runner，本章節，需要執行測試，要建立 Docker 類型的 Gitlab Runner。

先到 api 專案底下的 CICD - runner 底下，獲取該專案的 Token。

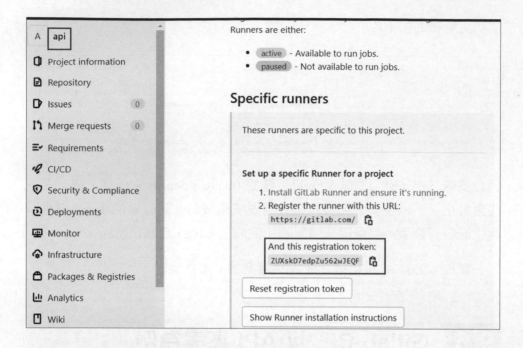

再執行以下指令：

```
sudo gitlab-runner register --non-interactive --url
"https://gitlab.com/" --registration-token
"ZUXskD7edpZu562wJEQF" --executor "docker" --docker-image
ubuntu:18.04 --description "docker-runner" --tag-list
"docker-runner"
```

這裡有幾點與註冊 shell runner 不同。

- --executor：指定在 Docker 環境下執行。
- --docker-image：指定 Docker 的 image。

執行後，等個 5 分鐘，看到以下畫面，就代表建立成功了。

有了專屬的 Runner 後，就可以針對資料庫設定，只對這個 Runner 所在的機器，開設防火牆，這樣可以同時達到，安全性與執行測試的條件了。

關於 .gitlab-ci.yml，寫法與 9.5 類似，如下，或可參考以下連結。

https://gitlab.com/FinMindBook/api/-/blob/feature/9.6.1/.gitlab-ci.yml

```
stages:
  # CI pipeline
  # 一般開發流程
  # test 測試過了 ->
  #   merge 進 stage/master ->
  #   build 建立 docker image ->
  #   deploy 部屬服務
  # 會根據此順序依序進行
  # test -> build -> deploy
```

```
  - test
  - build
  - deploy

# CI 名稱，test 步驟
test-api:
    # 在 stage 設定 CI pipeline 順序
    # 這裡 pipeline 步驟是 test
    stage: test
    # 由於使用專屬 Runner，所以不需要指定 image
    # image: ubuntu:18.04
    before_script:
      # 測試事前準備的指令
      # 需要先安裝環境
      - apt-get update && apt-get install python3.6 -y && apt-get install
python3-pip -y
      - pip3 install pipenv==2020.6.2
      - export LC_ALL=C.UTF-8 LANG=C.UTF-8
      - pipenv sync
      - python3 genenv.py
    script:
      # 實際測試指令
      pipenv run pytest --cov-report term-missing --cov-config=.
coveragerc --cov=./api/ tests/
    # tag，設定在跑測試時，使用本節建立的 runner 執行
    tags:
        - docker-runner
    only:
        # 設定只有在 merge requests 的情況下
        # 才會運作此 CI
        refs:
          - merge_requests
```

基本上指令都類似，Test 執行結果，也可到以下畫面查看：

https://gitlab.com/FinMindBook/api/-/merge_requests/1

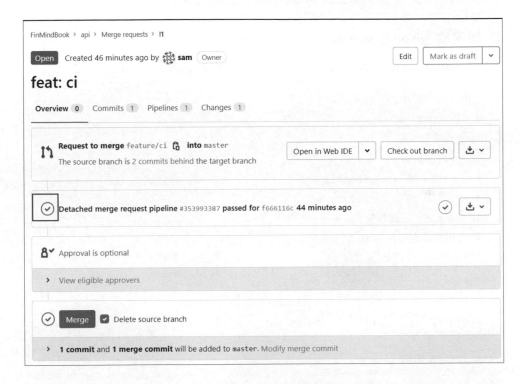

到這步驟，CI 自動化測試，已經完成，即使資料庫有設防火牆，只要對
該機器上的 Runner 開權限，一樣能執行 CI。下一章節，將介紹自動化建
立 Docker Image。

9.6.2 Build - 建立 Image

在這步驟，一樣要先建立 shell GitLab Runner，因為在 9.4 建立的，是專
門給 financialdata 專案使用，無法給 api 專案使用，因此使用同樣指令，
來建立 api 專屬的 shell GitLab Runner。

```
sudo gitlab-runner register --non-interactive --url "https://gitlab.
com/" --registration-token "ZUXskD7edpZu562wJEQF" --executor "shell"
--description "build_image" --tag-list "build_image"
```

CI 設定檔 .gitlab-ci.yml 如下：

https://gitlab.com/FinMindBook/api/-/blob/9.6.2/.gitlab-ci.yml

```
stages:
  # CI pipeline
  # 一般開發流程
  # test 測試過了 ->
  #   merge 進 stage/master ->
  #   build 建立 docker image ->
  #   deploy 部屬服務
  # 會根據此順序依序進行
  # test -> build -> deploy
  - test
  - build
  - deploy


# test-api 略過


# CI 名稱, 建立 docker image
build-docker-image:
    # 在 stage 設定 CI pipeline 順序
    # 這裡 pipeline 步驟是 build
    stage: build
    before_script:
        # 由於 build 好 image 後，需要 push 到 docker hub
        # 需要事先登入
        # 登入部分，這裡使用 token 做登入，可參考以下連結
```

```
    # https://docs.docker.com/docker-hub/access-tokens/#create-an-
access-token
    # token你可以使用明碼，或是存入gitlab CICD Variables，用變數的方式引用
    # 如下
    # docker login -u "linsamtw" -p 7777-7777-7777-7777
    # 但實務上，不希望 token 公開，因此會採用以下方式
    - docker login -u "linsamtw" -p ${DOCKER_HUB_TOKEN}
  script:
    # 建立 image
    - make build-image
    # push
    - make push-image
  tags:
    # 設定使用前面建立的 runner 執行
    - build_image
  only:
    # 設定只有在下 tag 後
    # 才會 build image
    - tags
```

在下圖步驟 1，新增 tag 後，確認下圖 3，CD Build 是成功的，就完成自
動化建立 Image 這步驟了。

接著到你的 Docker Hub 底下，就能看到，新版本 9.6.2 的 api Docker
Image，已經成功上傳了，下圖是筆者的 Docker Hub。

到這步驟，已經將測試、建立 image 自動化，一切都與爬蟲專案類似，甚至連 .gitlab-ci.yml 設定檔都沒做太多更動，以上都可以複製到讀者的專案中，讓你的專案也達到自動化。

下一章節，開始進行新版本部屬的自動化。

9.6.3 deploy - 部屬

先到 portainer，確認目前 api 版本，如下：

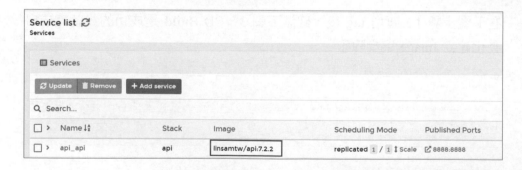

目前是 7.2.2 版本，等將自動化部屬做好後，只要新增 Tag，就會自動更新版本。

設定也與 9.5 相同，如下，也是在 CICD 中使用 Make 指令執行，將 GIT_
TAG 包進 Makefile 中，根據 Tag 更動 Docker Image 版本號。

.gitlab-ci.yml

https://gitlab.com/FinMindBook/api/-/blob/9.6.3/.gitlab-ci.yml

```
stages:
  # CI pipeline
  # 一般開發流程
  # test 測試過了 ->
  #   merge 進 stage/master ->
  #   build 建立 docker image ->
  #   deploy 部屬服務
  # 會根據此順序依序進行
  # test -> build -> deploy
  - test
  - build
  - deploy


# test-api 步驟省略


# CI 名稱, 建立 docker image
build-docker-image:
    # 在 stage 設定 CI pipeline 順序
    # 這裡 pipeline 步驟是 build
    stage: build
    before_script:
      # 由於 build 好 image 後，需要 push 到 docker hub
      # 需要事先登入
      # 登入部分，這裡使用 token 做登入，可參考以下連結
      # https://docs.docker.com/docker-hub/access-tokens/#create-an-
access-token
```

```
    # token你可以使用明碼,或是存入gitlab CICD Variables,用變數的方式引用
    # 如下
    # docker login -u "linsamtw" -p 7777-7777-7777-7777
    # 但實務上,不希望 token 公開,因此會採用以下方式
    - docker login -u "linsamtw" -p ${DOCKER_HUB_TOKEN}
  script:
    # 建立 image
    - make build-image
    # push
    - make push-image
  tags:
    # 設定使用前面建立的 runner 執行
    - build_image
  only:
    # 設定只有在下 tag 後
    # 才會 build image
    - tags


# CI 名稱, deploy api
deploy-api:
  # 在 stage 設定 CI pipeline 順序
  # 這裡 pipeline 步驟是 deply
  stage: deploy
  script:
    # 部屬 api
    - make deploy-api
  tags:
    # 設定使用前面建立的 runner 執行
    - build_image
  only:
    # 設定只有在下 tag 後
```

```
    # 才會 deploy
    - tags
```

Makefile 如下：

https://gitlab.com/FinMindBook/api/-/blob/9.6.3/Makefile

```
GIT_TAG := $(shell git describe --abbrev=0 --tags)

# 建立 docker image
build-image:
    docker build -f Dockerfile -t linsamtw/api:${GIT_TAG} .

# 推送 image
push-image:
    docker push linsamtw/api:${GIT_TAG}

# 部屬 api
deploy-api:
    GIT_TAG=${GIT_TAG} docker stack deploy --with-registry-auth -c api.
yml api
```

將以上設定，加入 master 後，新增 Tag 步驟如下。

到 api 專案底下，Repository -> Tags，New tag（新增一個 Tag），這邊新增的 Tag 是 9.6.3，且等到下圖 5，CD 運作結束後，再到 Portainer 畫面查看 api 的 Docker Image 是否有更新版本。

可到以下畫面，查看 CD 是否正常運作，以下為成功畫面，建立 Docker Image（build-docker-image），部屬新版本 api（deploy-api）。

以下為 Portainer 畫面，成功將 api 版本，更新到 9.6.3。

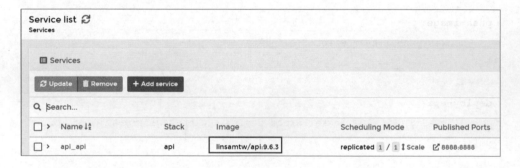

到這步驟後，關於 api 的自動化測試、建立 Docker Image、部屬新版本，都完成自動化了，讀者可以實際上體會看看，自動化的差別。單純建立新的 Tag，就會自動發布 Image 並部屬新版本，這對於後續的開發維運上，是非常方便的。

9.7 總結

第 9 章節的內容都完成後，已經達成，**爬蟲、API**，自動化測試（**CI**）、自動化部屬（**CD**），接著讀者只需要專注開發新爬蟲、新 API，剩下換版過程都定下來了，且使用分散式，因此也不用擔心效能問題。

特別的是自動化執行測試，可以確保在長期開發上專案的穩定性，不會發生團隊成員粗心大意，導致上線運作的程式，因為某次 Merge 後，掛掉無法正常運作。也避免發生團隊成員手動部屬新版本，粗心導致線上服務出問題。以上都是筆者親身經驗，因為缺少測試，導致新版本上線後，系統出現異常。

可能有讀者認為，手動也能正常運作，會出錯都是某些工程師的專業度不夠，或是刻意搞亂的。這句話是沒錯，但是，人工介入一定會有出錯的可能，而本書使用 CICD 建立了 SOP 流程，減少團隊開發上的成本，團隊只需專注在功能開發上，照著固定的流程做開發，用最有效率的方式做開發，這也是工程師的價值之一，不只是 Coding 而已，完善流程、制定架構會比單純寫程式來的重要。

第 3 篇
API 產品上線

Chapter

10

API 服務網址

10.1 為什麼需要網址？

本章節介紹如何設定 API 的服務網址，讓產品上線，那為什麼需要設定服務網址呢？有以下幾點好處：

第一，當需要提供其他人，讀者開發的 API 的服務時，相較於直接給對方 API 的 IP，用網址會相對好記憶，你不會希望連 google 網站，是用 142.251.42.227:443 這樣的 IP 去記憶，用以下網址跟 IP 相比，https://www.google.com.tw/，對一般人理解上，會更友善。

第二，SSL 安全性憑證。一般使用 IP 連上網時，都是 http，那要如何將 http 變成 https 呢？就需要 SSL 憑證，而這基本上需要有網址才能做設定，SSL 有什麼好處？這是現代標準的加密技術，一般進行網路瀏覽，可能會傳送一些個資，如信用卡資訊，那這中途，是有可能被攔截、擷取特定資訊，因此才需要經過加密階段。即使不考慮這些相對複雜的問題，如果不用 SSL，對於你的網站在搜尋引擎上，是會被扣分的，甚至不讓你出現在搜尋列表中。

第三，搜尋引擎。當你使用 IP 做為網站時，搜尋引擎是不會收錄的，也就是說，使用者無法透過搜尋，找到你的 API 或是網站。

實際上用網址，甚至 DNS，都有更多好處，例如 Email、子網域設定、Loading Balance 等等，就不在此多做介紹，初期只要知道，用網址好記憶、SSL 安全性、搜尋引擎相對高分，這幾點就夠了。

10.2 No-Ip 免費的網址申請

一般網址都需要花錢購買，但多數讀者沒有商用，單純在 Side Project 階段，因此盡可能使用免費服務，以下將介紹 No-IP 這個免費的網址申請網站。

https://www.noip.com/

先用以上網址註冊並登入後，到以下網址：

https://my.noip.com/

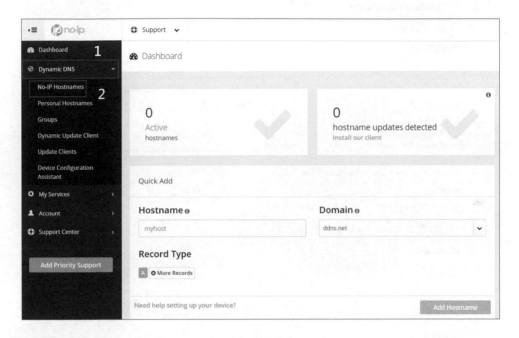

點選 **Dynamic DNS** 後，再點選 **No-IP Hostnames**。

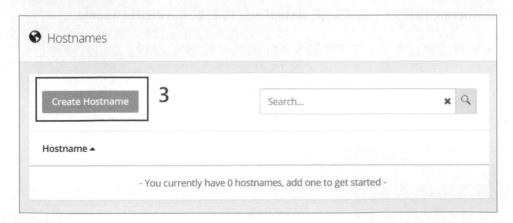

接著點擊 **Create Hostname**,建立網址。

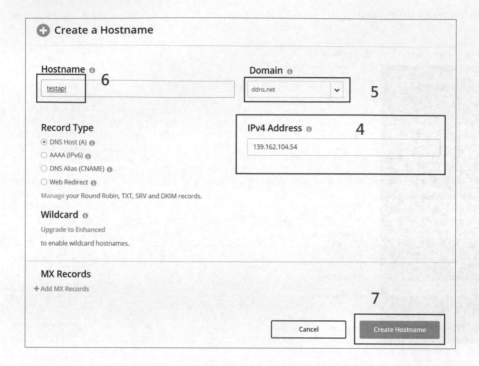

在 **IPv4 Address** 輸入你的 Linode 的 IP,**Domain** 隨便選一個你喜歡的,**Hostname** 也自訂,這裡輸入 **testapi**,最後點擊 **Create Hostname**。

接著就設定完成了,但這時因為 API 的 Port 設定為 8888,無法直接連上。

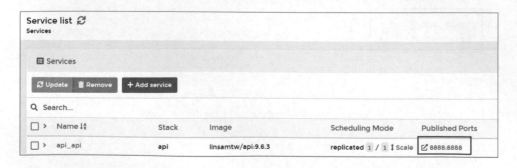

因為一般網站連線，預設都是連 80 Port（http）或是 443 Port（https），
所以到 portainer 上，將 api 的 Port 改為 80。

先到 Services > api_api 底下，點選 Network & published ports：

將 Host port 改為 80：

改完後，點選 Apply changes：

這時再到剛剛設定的網址，http://testapi.ddns.net/docs，就能看到 API 了，是不是很簡單？

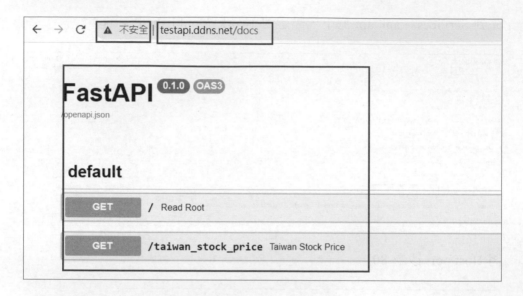

這裡注意一點，在圖上 testapi.ddns.net/docs 前面，有個三角形驚嘆號，顯示不安全，因為目前使用的是 http，下一章節，將會講解如何設定 SSL。

10.3　Let's Encrypt 免費的 SSL 憑證

同樣，SSL 也有免費與付費版本，本書介紹 Let's Encrypt，免費的 SSL 憑證，一般常見的是使用 Nginx 作設定，進階會搭配 HAProxy 做 Loading Balance，但現代出現 Swarm、K8s 等集群容器架構後，衍生出相對應的工具，本書介紹 **Traefik**，能更好的管理所有服務的網址、DNS，並自動更新 SSL，且有不錯的 Dashboard 工具，監控所有網址狀態，如下圖。

以下將介紹，如何將 API 結合 Traefik，自動完成 SSL，將你的 API 網址，加入 https。

10.4 Traefik

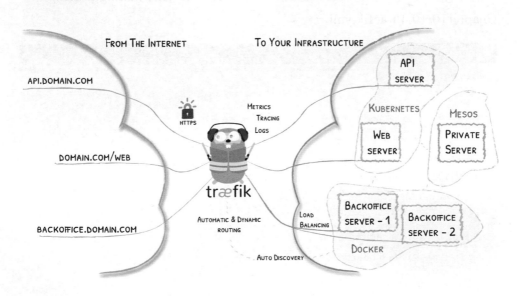

以上圖片引用 Traefik 官方網站，https://doc.traefik.io/traefik/。

中間是 Traefik 圖示，代表在 Swarm 中，Manager 的機器。Traefik 左邊是讀者設定的各種網址，想像一個情境，團隊中，開發的產品包含 API、Web、Backend，以上三個都需要相對應的網址，如 api.com、web.com、backend.com，而 Traefik 右邊，則是是相對應的服務，api_server、backend_server、web_server。

一般使用者，會藉由不同的網址，拜訪你的網站、API，如左圖，經過 Traefik 後，Traefik 會根據不同網址，回傳不同的服務。如果拜訪的網址是 api.com，Traefik 就會回應 api_server，如果拜訪的是 web.com，Traefik 就會回應 web_server。基本上是由 Traefik 統一管理，使用者全部都從同一個入口，Traefik，拜訪你的服務，這跟前面提的 Manager 概念很像，現代架構，都是講求統一管理，以下將講解如何實作。

首先，先架設 Traefik，筆者一樣使用 Docker 架設，設定檔如下：

https://github.com/FinMind/FinMindBook/blob/master/DataEngineering/Chapter10/10.4/traefik.yml

```
traefik.yml

version: "3.8"


services:

  traefik:
    image: "traefik:v2.2" # image 版本
    command:
      # 設定顯示 dashboard
      - --api.insecure=true
      - --api.dashboard=true
```

```
    # log 模式，有 DEBUG、ERROR、INFO 等模式
    - --api.debug=true
    - --log.level=ERROR
    # https://doc.traefik.io/traefik/v2.2/routing/providers/docker/
    # 根據官方文件，有 docker、k8s 等模式可做選擇
    # 在此選擇 docker
    - --providers.docker=true
    - --providers.docker.endpoint=unix:///var/run/docker.sock
    - --providers.docker.swarmMode=true
    - --providers.docker.exposedbydefault=false
    - --providers.docker.network=traefik-public
    # 進入點，80、443 分別對應 http、https
    - --entrypoints.web.address=:80
    - --entrypoints.web-secured.address=:443
    # SSL 設定
    - "--certificatesresolvers.myresolver.acme.httpchallenge=true"
    - "--certificatesresolvers.myresolver.acme.httpchallenge.
entrypoint=web"
    - "--certificatesresolvers.myresolver.acme.email=
samlin266118@gmail.com"
    - "--certificatesresolvers.myresolver.acme.storage=/letsencrypt/
acme.json"
  ports:
    # 開對外 port，除了 80、443 是給 http、https 外
    # 8080 port 給 dashboard
    # 但 8080 已經給 phpmyadmin
    # 因此這裡 published 到 8888
    - target: 80
      published: 80
      mode: host
    - target: 443
```

```yaml
          published: 443
          mode: host
        - target: 8080
          published: 8888
          mode: host
      volumes:
        # 存放 SSL 憑證
        - "letsencrypt:/letsencrypt"
        - "/var/run/docker.sock:/var/run/docker.sock:ro"
      restart: unless-stopped
      # log 最大 size
      logging:
          driver: "json-file"
          options:
              max-size: "50m"
      # deploy 設定
      deploy:
        replicas: 1
        update_config:
          parallelism: 1
          delay: 10s
          order: stop-first
          failure_action: rollback
        placement:
          max_replicas_per_node: 1
          constraints: [node.role == manager]
      networks:
          - traefik-public

networks:
  traefik-public:
    external: true
```

```
volumes:
  letsencrypt:
```

本書使用 Traefik 2.2 版本，版本 1 跟 2 存在蠻大的差異。

Log 模式選用 ERROR，如果在測試階段，可以選用 DEBUG 或是 INFO，觀察 Traefik 的 Log。Providers 選用 Docker，entrypoints 進入點設定 80 與 443，分別是 http 與 https。另外需要將 SSL 中的 Email 改成讀者自己的，而 ACME 證書存放在 /letsencrypt/acme.json，其餘就是 Swarm 的一些設定，由於前面章節都介紹過了，相信到了第十章節，讀者也很熟悉了，以下直接進行部屬。

由於 Traefik 要使用 80 Port，因為先前已將 80 Port 用於 api，在此先進行回收，如下，關閉 api 的 Published Ports：

☐ >	Name ↓↑	Stack	Image	Scheduling Mode	Published Ports
☐ >	api_api	api	linsamtw/api:10	replicated 1 / 1 ⑂ Scale	-

接著建立 Traefik 專屬的 Network：

```
docker network create --driver=overlay traefik-public
```

再部屬 Traefik：

```
docker stack deploy -c traefik.yml tr
```

部屬完成後，到 http://139.162.104.54:8888/dashboard/#/，會看到以下畫面：

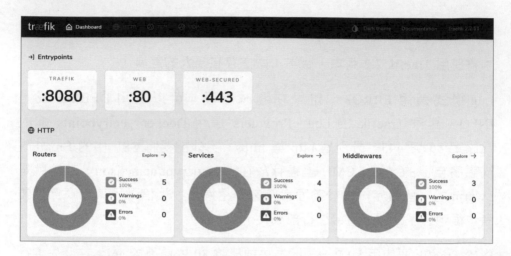

Traefik 好處是有 UI 介面，讀者可以試著在上面先看看各種功能，相較於 Nginx，Traefik 與其他工具有更好的界接，如常見的監控工具 Promethus、Grafana，以上兩個服務也可以使用 Docker 進行架設，本書就不多做介紹，讀者可自行研究。

架設完 Traefik 後，API 需要做額外設定，才能跟 Traefik 結合，別擔心，本書也會一一介紹如何設定。

10.5　API 結合 Traefik

前一個章節，介紹 Traefik，統一了 User 的訪問入口，因此 API 不需要把 Port 串出去，取而代之的是，在 API 的 docker-compose.yml 做設定，讓 Traefik 認得 API，當使用者拜訪 http://testapi.ddns.net/docs（前面申請的網址），Traefik 會自動導向 API 這個 Server，設定檔如下：

https://github.com/FinMind/FinMindBook/blob/master/DataEngineering/
Chapter10/10.5/api/api.yml

```yml
api.yml
version: '3.0'
services:
  api:
    image: linsamtw/api:10
    hostname: "api"
    restart: always
    # swarm 設定
    deploy:
      mode: replicated
      replicas: 1
      placement:
        constraints: [node.labels.api == true]
      labels:
        - traefik.enable=true
        - traefik.frontend.passHostHeader=true
        # routers 設定,
        # 這裡根據 fastapi 所需的 router 做額外設定
        # fastapi 預設在 /docs 作為文件頁面,
        # 並根據 /openapi.json 去製作文件
        # 因此需額外設定
        # testapi.ddns.net/docs 與
        # testapi.ddns.net/openapi.json
        # 可被 traefik 訪問,方法如下
        # 指定 Host 與 Path
        # &&:代表 and,將 Host 與 Path 連在一起使用
        # ||:代表 or,在多個 router 時使用
        - traefik.http.routers.api-https.rule=
          Host(`testapi.ddns.net`) && PathPrefix(`/docs`) ||
```

```
        Host(`testapi.ddns.net`) && PathPrefix(`/openapi.json`) ||
        Host(`testapi.ddns.net`)
      # api loading balance 的 port
      # traefik 的好處之一，會幫你處理分流
      - traefik.http.services.api-https.loadbalancer.server.port=80
      # SSL 設定
      - traefik.http.routers.api-https.tls.certresolver=myresolver
      # http 自動導向 https
      - "traefik.http.routers.api-http-catchall.rule=hostregexp
(`{host:.+}`)"
      - "traefik.http.routers.api-http-catchall.entrypoints=web"
      - "traefik.http.routers.api-http-catchall.middlewares=api-
redirect-to-https@docker"
      - "traefik.http.middlewares.api-redirect-to-https.redirectscheme.
scheme=https"
    environment:
      - TZ=Asia/Taipei
    networks:
      - my_network
      - traefik-public

networks:
  my_network:
    # 加入已經存在的網路
    external: true
  traefik-public:
    # 需加入 traefik 的網路
    external: true
```

以上跟 8.1.5 的 api.yml 相比，移除 Port，因為改走 Traefik。多了 labels 的設定，主要都是跟 Traefik 有關，其中：

traefik.enable=true：啟動 Traefik。

traefik.frontend.passHostHeader=true：傳送 Host Header 給 API Server。想像一個情境，如果 API 會記錄每一個 Request 的資訊，如 IP，但當使用 Traefik 架構後，所有 User 都是訪問 Traefik，藉由 Traefik 轉發 Request 到 API，因此這時 API 得到的 User IP，全部會變成 Traefik 的 IP，但這並不是實際上 User 的 IP，因此需要加入此設定，才能獲得 User 真實的 IP。

traefik.http.routers.api-https.rule：設定 Router 規則。Host 是前面使用 No-IP 申請的網址 testapi.ddns.net，Path 可以看成 API 的路徑。舉例來說，FastAPI 自動生成的文件，會在 /docs 路徑底下，因此需要將此規則（Host(`testapi.ddns.net`) && PathPrefix(`/docs`)）， 加 到 Router 中，Traefik 才有辦法辨別。同理，如果讀者對於 API 有額外的版本設定，如 /api/v1、/api/v2，那 /api 這層 path，就需要加入此，Host(`testapi.ddns.net`) && PathPrefix(`/api`)。

traefik.http.services.api-https.loadbalancer.server.port=80：Traefik 的好處之一，會自動處理分流（Loading Balance），此 Port 設定為 80。

另一個不同的地方是，需要加入 traefik-public 的 Network，在同一個網路底下，才能互通，這也是 Docker 的特色之一。除了 api.yml 需要寫額外設定，因為 Traefik Loading Balance 是指定 80 Port，需要將 Dockerfile 中的 Port 改為 80，如下：

https://github.com/FinMind/FinMindBook/blob/master/DataEngineering/Chapter10/10.5/api/Dockerfile

```
Dockerfile
FROM continuumio/miniconda3:4.3.27

RUN apt-get update

RUN mkdir /FinMindProject
COPY . /FinMindProject/
WORKDIR /FinMindProject/

# install package
RUN pip install pipenv && pipenv sync

# genenv
RUN VERSION=RELEASE python genenv.py

# 預設執行的指令
CMD ["pipenv", "run", "uvicorn", "api.main:app", "--host", "0.0.0.0",
"--port", "80"]
```

到這就結束了 API 與 Traefik 的結合,實際上並沒有做太多改動,但多了 SSL、自動更新 SSL 憑證、Loading Balance、Router 管控,接著開始將 以上新版的 api.yml 部屬上線,指令如下:

重新建立 Docker Image,linsamtw/api:10

```
docker build -f Dockerfile -t linsamtw/api:10 .
```

push

```
docker push linsamtw/api:10
```

部屬

```
docker stack deploy -c api.yml api
```

完成以上指令後,到以下網址:

http://139.162.104.54:8888/dashboard/#/http/routers

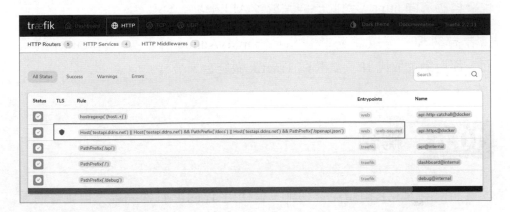

如上圖,可以觀察到,增加了新的 Rule,Host(`testapi.ddns.net`) || Host
(`testapi.ddns.net`) && PathPrefix(`/docs`) || Host(`testapi.ddns.net`) &&
PathPrefix(`/openapi.json`),這就是前面在 api.yml 的設定,注意一點,在
TLS 有個綠色盾牌,代表成功加入 SSL 憑證,這時再回到 API 的網址:

http://testapi.ddns.net/docs

如上,從三角形驚嘆號,變成鑰鎖了,代表成功加入 SSL,而 API 的網址,也變成 **https** 了,即使輸入 http 的網址,http://testapi.ddns.net/docs,也會自動導向 https,到此已經完成了 API 網址、SSL。

讀者有興趣,可以試著將資料庫 phpMyAdmin 的頁面、Portainer 的頁面,都用網址取代 IP,試著與 Traefik 做結合。

10.6 總結

本章節使用 No-IP 申請免費網址,並使用 Traefik 管理網址、自動註冊更新 SSL 憑證,這樣對於提供 API 產品,又更進一步,實務上基本都是用網址、甚至 DNS,不太可能直接提供 IP,這對於工作或是求職上,更進一步加深技能樹。

筆者初期也是使用 Nginx 處理網址、用 Certbot 自動獲取 Let's Encrypt 憑證,但後期接觸 Docker 後,將一切都容器化,當然 Nginx 也可以用 Docker,但與專門用於集群架構的 Traefik 相比,Traefik 在管理與開發上,非常容易上手,且高拓展性,因此本書特別介紹 Traefik。讀者在 API 開發的路上,一定會遇到網址、SSL 的問題,使用 Traefik 會讓一切都變的美好。

第 4 篇
資料視覺化

11.1　什麼是視覺化？

在大數據的時代，要怎麼使用大量數據，協助商業分析、數據分析，導入數據驅動（Data Driver）的觀念，是一個蠻大的課題。以本書金融資料為例，到了第 11 章，讀者已經完成爬蟲，收集了股價資料，只要持續開發各種爬蟲，就能收集超過 10 種金融資料，收集完成後，要如何使用這些資料呢？

視覺化是一個蠻常被提起的概念，且在業界，視覺化報表是非常通用的工具，基本上有使用數據的公司，一定會導入 Dashboarsd，可能用於業務分析、流量分析、營運分析、KPI 等等，那要如何製作 Dashboard 呢？

Python 常見的繪圖工具有 matplotlib、plotly 甚至 dash，但這些都是偏向單機版，多數用於 Jupyter 上，且要寫不少程式，那有沒有更簡單的工具？可以更好的呈現給業務單位、行銷單位觀察數據呢？Redash 就是現成的 BI（Business Intelligence）工具，下圖就是筆者使用 Redash 製作的一個範例。

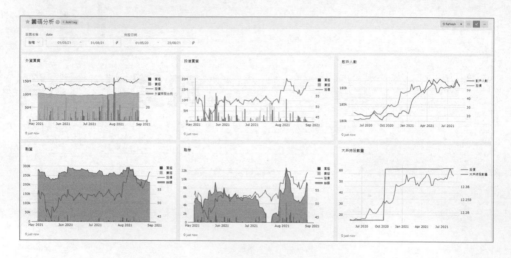

上圖單純使用 Docker 架設好服務後，剩下只需要寫 SQL 即可做出 Dashboard，相較於寫 Python，開發速度快非常多，且 SQL 相對簡單，可以提供更多業務端、分析端使用，不再限於工程團隊。

可能有些讀者比較想做模型，而非視覺化，但實務上，並沒有這麼單純，不論是做模型或是資料分析，最後的結果，都需要呈現給其它部門，使用 Data 協助他們解決業務上的問題，而最好的呈現方式，就是使用 BI 工具製作 Dashboard。

且實務上並不會所有 Data 相關的問題，都需要模型，例如營運單位希望觀察每天的業績、網站流量，行銷部門需要觀察，針對不同管道投放廣告的差異，以上都會是 Data 部門負責，且都跟模型無關，這時最好的方式，就是製作 Dashboard 作呈現，以下將從架設 Redash 開始一一講解。

11.2   Redash

相信讀者看到第 11 章節後，都能猜到本書要用什麼方式去架設 Redash 了，沒錯，就是 Docker，只不過 Redash 使用的相關服務稍微多一點，根 據 Redash 官 方 GitHub，https://github.com/getredash/redash， 啟 動 Redash，至少就需要 Postgres（RDB 資料庫）、Redis（NoSQL 資料庫）、Nginx（網頁伺服器）、Worker（分散式）這四個相對應的服務，還不包含 Redash Server 本身，這也是為什麼 Redash 是業界常用的視覺化工具，因為 Redash 將各個部分，拆解成微服務 Docker，並使用 Worker 分散式、使用 Redis 做圖表暫存，是很強大的 BI 工具。

這裡突然出現幾個名詞，讀者如果第一次看到，可能有點陌生，筆者在此做一些簡單介紹。

Postgres，類似於 MySQL 的 RDB 資料庫，都是關聯式資料庫，主要是使用場景不同，初學者可以把它跟 MySQL 看成一樣的東西，等遇到不同的情境，再比較差異，基本上兩者都有不同的擁護者。

Redis，NoSQL 資料庫，最大的優勢是，資料存放在記憶體層級，使它反應非常快速，由於網站對於反應時間有一定的要求，越快越好，因此這裡 Redash 提供 Web 服務，一部分就是使用 Redis 存放暫存檔。

以上介紹完後，筆者開始講解如何架設 Redash，yml 檔如下：

https://github.com/FinMind/FinMindBook/blob/master/DataEngineering/
Chapter11/11.2/redash.yml

```
redash.yml

version: "3.8"
services:
```

```
redash:
  # 選用此版本的 redash image
  image: redash/redash:8.0.0.b32245
  # depends 代表依賴關係
  # 需要等 postgres、redis 先架設完成
  # 才會開始架設 redash services
  depends_on:
    - postgres
    - redis
  restart: always
  # redash 啟動 server 指令
  command: server
  # redash 使用 5000 port
  ports:
    - target: 5000
      published: 5000
      mode: host
  # 相關的環境變數,包含
  environment:
    PYTHONUNBUFFERED: 0
    # log 等級
    REDASH_LOG_LEVEL: INFO
    # redis 的連線設定
    REDASH_REDIS_URL: redis://redis:6379/0
    # postgres 的密碼
    POSTGRES_PASSWORD: postgres_password
    REDASH_COOKIE_SECRET: redash_cookie_secret
    REDASH_SECRET_KEY: redash_secret_key
    # postgres 的連線設定
    REDASH_DATABASE_URL: postgresql://postgres:postgres_password@postgres/
postgres
    # 允許在 redash 上使用 python
```

```
      REDASH_ADDITIONAL_QUERY_RUNNERS: redash.query_runner.python
      # worker 數量
      # 這是做範例，設定 1，
      # 未來可根據需求自行增加
      REDASH_WEB_WORKERS: 1
    networks:
      - my_network

  create_table:
    # 相關設定與 redash services 相同
    # 也用相同的 image
    image: redash/redash:8.0.0.b32245
    depends_on:
      - postgres
      - redis
    # restart: always
    # 主要差異在此
    # 初始化需要建立 table
    # 建立完成後，可在 portainer 上刪除此 services
    command: python /app/manage.py database create_tables
    environment:
      PYTHONUNBUFFERED: 0
      REDASH_LOG_LEVEL: INFO
      REDASH_REDIS_URL: redis://redis:6379/0
      POSTGRES_PASSWORD: postgres_password
      REDASH_COOKIE_SECRET: redash_cookie_secret
      REDASH_SECRET_KEY: redash_secret_key
      REDASH_DATABASE_URL: postgresql://postgres:postgres_password@postgres/
postgres
      REDASH_ADDITIONAL_QUERY_RUNNERS: redash.query_runner.python
      REDASH_WEB_WORKERS: 1
    networks:
```

```
      - my_network

  scheduler:
    # 用於 redash 相關的 scheduler 排程
    image: redash/redash:8.0.0.b32245
    depends_on:
      - postgres
      - redis
    restart: always
    command: scheduler
    environment:
      PYTHONUNBUFFERED: 0
      REDASH_LOG_LEVEL: INFO
      REDASH_REDIS_URL: redis://redis:6379/0
      POSTGRES_PASSWORD: postgres_password
      REDASH_COOKIE_SECRET: redash_cookie_secret
      REDASH_SECRET_KEY: redash_secret_key
      REDASH_DATABASE_URL: postgresql://postgres:postgres_password@postgres/
postgres
      REDASH_ADDITIONAL_QUERY_RUNNERS: redash.query_runner.python
      # 這裡的 celery 就跟前面章節提到的
      # 分散式概念一樣
      QUEUES: "celery"
      WORKERS_COUNT: 1
    networks:
        - my_network

  scheduled_worker:
    # 分散式中的 worker
    image: redash/redash:8.0.0.b32245
    depends_on:
      - postgres
```

```
    - redis
  restart: always
  command: worker
  environment:
    PYTHONUNBUFFERED: 0
    REDASH_LOG_LEVEL: INFO
    REDASH_REDIS_URL: redis://redis:6379/0
    POSTGRES_PASSWORD: postgres_password
    REDASH_COOKIE_SECRET: redash_cookie_secret
    REDASH_SECRET_KEY: redash_secret_key
    REDASH_DATABASE_URL: postgresql://postgres:postgres_password@postgres/
postgres
    REDASH_ADDITIONAL_QUERY_RUNNERS: redash.query_runner.python
    # 負責處理 queue = scheduled_queries,schemas 類型的任務
    # 與本書中 7.6.3，crawler.yml 中的 -Q twse 概念一樣
    QUEUES: "scheduled_queries,schemas"
    WORKERS_COUNT: 1
  networks:
      - my_network

adhoc_worker:
  image: redash/redash:8.0.0.b32245
  depends_on:
    - postgres
    - redis
  restart: always
  command: worker
  environment:
    PYTHONUNBUFFERED: 0
    REDASH_LOG_LEVEL: INFO
    REDASH_REDIS_URL: redis://redis:6379/0
    POSTGRES_PASSWORD: postgres_password
```

```
    REDASH_COOKIE_SECRET: redash_cookie_secret
    REDASH_SECRET_KEY: redash_secret_key
    REDASH_DATABASE_URL: postgresql://postgres:postgres_password@postgres/
postgres
    REDASH_ADDITIONAL_QUERY_RUNNERS: redash.query_runner.python
    # 負責處理 queue = queries 類型的任務
    # 與本書中 7.6.3，crawler.yml 中的 -Q twse 概念一樣
    QUEUES: "queries"
    WORKERS_COUNT: 2
  networks:
    - my_network

 redis:
  # redis 資料庫
  image: redis:5.0
  restart: always
  volumes:
    - redis-redash:/bitnami/redis/data
  networks:
    - my_network

 postgres:
  # postgres 資料庫
  image: postgres:9.6-alpine
  environment:
    PYTHONUNBUFFERED: 0
    REDASH_LOG_LEVEL: INFO
    REDASH_REDIS_URL: redis://redis:6379/0
    POSTGRES_PASSWORD: postgres_password
    REDASH_COOKIE_SECRET: redash_cookie_secret
    REDASH_SECRET_KEY: redash_secret_key
    REDASH_DATABASE_URL: postgresql://postgres:postgres_password@postgres/
```

```
postgres
      REDASH_ADDITIONAL_QUERY_RUNNERS: redash.query_runner.python
    volumes:
      - postgres:/var/lib/postgresql/data
    restart: always
    networks:
        - my_network

  nginx:
    # 網頁伺服器
    image: redash/nginx:latest
    depends_on:
      - redash
    restart: always
    networks:
        - my_network

networks:
  my_network:
    external: true

volumes:
  postgres:

  redis-redash:
```

在部屬之前，有一點需要注意，一開始本書使用的雲端 Linode，是最低方案，只有 1GB ram，由於 Redash 需要多個服務，因此麻煩讀者將機器升級到 4GB ram 方案，20 美金／月（如果是使用個人電腦的讀者，則無此問題）。

部屬指令如下：

```
docker stack deploy -c redash.yml redash
```

成功執行後，畫面如下，建立了 7 個 Service：

```
sam@localhost:~/FinMindBook/DataEngineering/Chapter11/11.2$ docker stack deploy -c redash.yml redash
Ignoring unsupported options: restart

Creating service redash_scheduler
Creating service redash_scheduled_worker
Creating service redash_adhoc_worker
Creating service redash_redis
Creating service redash_postgres
Creating service redash_nginx
Creating service redash_redash
Creating service redash_create_table
sam@localhost:~/FinMindBook/DataEngineering/Chapter11/11.2$
```

接著轉到 Portainer 的頁面，如下圖，在 1 的地方，輸入 redash 做 filter，
之後在 2 的地方，可以看到前面建立的 redash 服務，Port 是 5000：

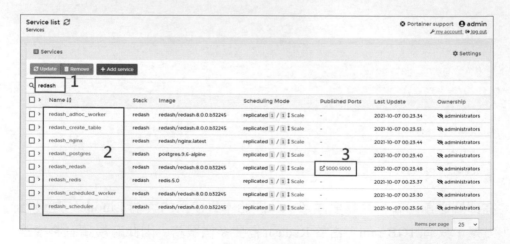

接著到以下網址 http://139.162.104.54:5000（IP 請自行換成讀者的 Linode
網址），如果看到以下畫面，代表部屬成功：

這時就可以把 create_table 的 service 移除了，方法如下：

這時就完成了 Redash 的部屬了，是不是很簡單，讀者可搭配 Chapter 10 的 Traefik，幫你的 Redash 申請網址，下一章節，將講解如何使用 Redash。

11.3 Redash 帳號設定

首先，先進行帳號設定，如下圖：

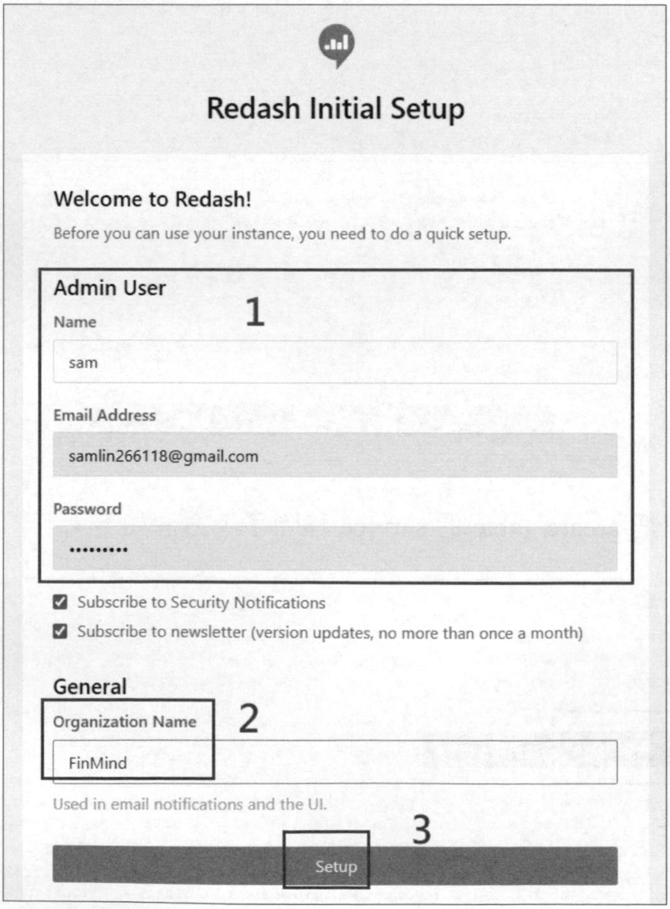

在 1 的地方，輸入你的名稱、Email、密碼，並在 2 的地方，輸入你的組織名稱，隨便你喜歡輸入什麼都可以，最後在 3 點選 Setup。接著就會進入以下畫面：

如果想幫團隊成員，開設帳號的話，可以按照以下步驟，點選右上角自己的圖示，選取 Users，最後到 3 的地方，New User，就可以新增使用者啦，非常的方便。

下一章，介紹如何與 MySQL 資料庫做連接，拿取金融資料做視覺化。

11.4 資料庫連接

如下圖，點選右上角圖示，再選取 Data Sources：

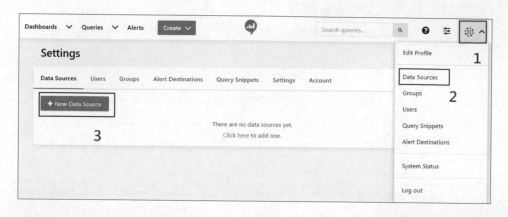

到以下畫面後，可以看到有超過 20 種的 DataBase 可以選擇，包含知名的 BigQuery，有興趣讀者可再自行研究，這裡筆者使用 MySQL，在 1 的地方輸入 MySQL 後，點選 MySQL：

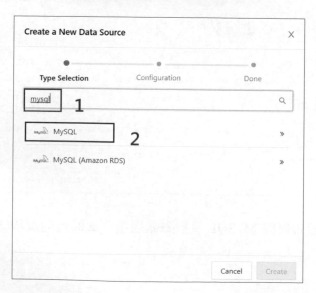

會再看到以下畫面，就一一輸入各個設定，Name、Host、Port、User、Password、Database name，特別注意一點，Host 的地方，一般來說是輸入 IP，但因為本書中在 Swarm 架構下，且跟 MySQL 統一個 Network，因此 Host 輸入 Service Name - mysql，是可以的。

11.5 匯入資料

在製作視覺化儀表板之前，麻煩讀者先自行收集，台股股價、三大法人買賣、融資融券、外資持股等資料，有資料才有辦法做視覺化。收集方法也很簡單，只要將章節 8.1.4 的部分，照著模仿股價爬蟲，因為證交所爬蟲架構都非常類似，可以輕易地複製出三大法人、融資融券、外資持股爬蟲等。

由於資料收集會花費許多時間，本書先將所需資料，上傳雲端，並提供程式碼，讓讀者可以直接將資料上傳到讀者架設的資料庫中，如下：

https://github.com/FinMind/FinMindBook/blob/master/DataEngineering/Chapter11/11.5/upload_data2mysql.py

```
upload_data2mysql.py
import os
import sys

import pandas as pd
from loguru import logger
from sqlalchemy import (
    create_engine,
    engine,
)
from tqdm import tqdm

import wget


def get_mysql_financialdata_conn() -> engine.base.Connection:
    # TODO 請將 IP 換成讀者自己的 IP
    address = "mysql+pymysql://root:test@139.162.104.54:3306/financialdata"
```

```
    engine = create_engine(address)
    connect = engine.connect()
    return connect



def create_taiwan_stock_info_sql():
    return """
        CREATE TABLE `taiwan_stock_info` (
            `industry_category` varchar(32) CHARACTER SET utf8 COLLATE
utf8_unicode_ci NOT NULL,
            `stock_id` varchar(32) CHARACTER SET utf8 COLLATE utf8_
unicode_ci NOT NULL,
            `stock_name` varchar(30) CHARACTER SET utf8 COLLATE utf8_
unicode_ci DEFAULT NULL,
            `type` varchar(4) CHARACTER SET utf8 COLLATE utf8_unicode_ci
NOT NULL COMMENT '上市twse/上櫃tpex',
            `date` date DEFAULT NULL,
            PRIMARY KEY (`stock_id`,`industry_category`)
        ) ENGINE=InnoDB DEFAULT CHARSET=utf8mb3 COLLATE=utf8_unicode_ci;
    """



def create_taiwan_stock_price_sql():
    return """
        CREATE TABLE `taiwan_stock_price` (
            `StockID` varchar(10) COLLATE utf8_unicode_ci NOT NULL,
            `Transaction` bigint NOT NULL,
            `TradeVolume` int NOT NULL,
            `TradeValue` bigint NOT NULL,
            `Open` float NOT NULL,
            `Max` float NOT NULL,
            `Min` float NOT NULL,
```

```
        `Close` float NOT NULL,
        `Change` float NOT NULL,
        `Date` date NOT NULL,
        PRIMARY KEY(`StockID`, `Date`)
    ) ENGINE=InnoDB DEFAULT CHARSET=utf8mb3 COLLATE=utf8_unicode_ci
    PARTITION BY KEY (StockID)
    PARTITIONS 10;
    """


def create_taiwan_stock_institutional_investors_sql():
    return """
        CREATE TABLE taiwan_stock_institutional_investors(
            `name` VARCHAR(20),
            `buy` BIGINT(64),
            `sell` BIGINT(64),
            `stock_id` VARCHAR(10),
            `date` DATE,
            PRIMARY KEY(`stock_id`, `date`, `name`)
        ) PARTITION BY KEY(`stock_id`) PARTITIONS 10;
    """


def create_taiwan_stock_margin_purchase_short_sale_sql():
    return """
        CREATE TABLE `taiwan_stock_margin_purchase_short_sale`(
            `stock_id` VARCHAR(10) CHARACTER SET utf8 COLLATE utf8_
unicode_ci NOT NULL COMMENT '股票代碼',
            `MarginPurchaseBuy` BIGINT NOT NULL COMMENT '融資買進',
            `MarginPurchaseSell` BIGINT NOT NULL COMMENT '融資賣出',
            `MarginPurchaseCashRepayment` BIGINT NOT NULL COMMENT
'融資現金償還',
```

```
        `MarginPurchaseYesterdayBalance` BIGINT NOT NULL COMMENT
'融資昨日餘額',
        `MarginPurchaseTodayBalance` BIGINT NOT NULL COMMENT
'融資今日餘額',
        `MarginPurchaseLimit` BIGINT NOT NULL COMMENT '融資限額',
        `ShortSaleBuy` BIGINT NOT NULL COMMENT '融券買進',
        `ShortSaleSell` BIGINT NOT NULL COMMENT '融券賣出',
        `ShortSaleCashRepayment` BIGINT NOT NULL COMMENT '融券償還',
        `ShortSaleYesterdayBalance` BIGINT NOT NULL COMMENT '融券昨日
餘額',
        `ShortSaleTodayBalance` BIGINT NOT NULL COMMENT '融券今日餘額',
        `ShortSaleLimit` BIGINT NOT NULL COMMENT '融券限制',
        `OffsetLoanAndShort` BIGINT DEFAULT NULL COMMENT '資券互抵',
        `date` DATE NOT NULL COMMENT '日期',
        PRIMARY KEY(`stock_id`, `date`)
    ) PARTITION BY KEY(`stock_id`) PARTITIONS 10;
    """


def create_taiwan_stock_holding_shares_per_sql():
    return """
    CREATE TABLE taiwan_stock_holding_shares_per (
        `HoldingSharesLevel` VARCHAR(19),
        `people` INT(10),
        `unit` BIGINT(64),
        `percent` FLOAT,
        `stock_id` VARCHAR(10),
        `date` DATE,
        `update_time` DATETIME NOT NULL DEFAULT CURRENT_TIMESTAMP,
        PRIMARY KEY (`stock_id`,`date`,`HoldingSharesLevel`)
    )
    PARTITION BY KEY(stock_id)
```

```
        PARTITIONS 10;
    """


def create_table(table: str):
    mysql_conn = (
        get_mysql_financialdata_conn()
    )
    sql = eval(f"create_{table}_sql()")
    try:
        logger.info(
            f"create table {table}"
        )
        mysql_conn.execute(sql)
    except:
        logger.info(
            f"{table} already exists"
        )


def download_data(table: str):
    logger.info("download data")
    if f"{table}.csv" in os.listdir(
        "."
    ):
        logger.info(f"already download")
    else:
        url = f"https://github.com/FinMind/FinMindBook/releases/download/
data/{table}.csv"
        wget.download(
            url, f"{table}.csv"
        )
```

```python
        logger.info(
            "download data complete"
        )



def upload_data2mysql(table: str):
    chunk_size = 100000
    mysql_conn = (
        get_mysql_financialdata_conn()
    )
    try:
        logger.info("load data")
        logger.info("upload to mysql")
        reader = pd.read_csv(
            f"{table}.csv",
            chunksize=chunk_size,
        )
        for df_chunk in tqdm(reader):
            df_chunk.to_sql(
                name=table,
                con=mysql_conn,
                if_exists="append",
                index=False,
            )
    except:
        logger.info("already upload")



def main(table: str):
    create_table(
        table=table,
    )
```

```
    download_data(
        table=table,
    )
    upload_data2mysql(table=table)


if __name__ == "__main__":
    table = sys.argv[1]
    main(table)
```

其中需注意一點，把 MySQL 的 IP，139.162.104.54，換成讀者自己的，才能成功上傳資料。

以上包含台股基本資料、台股股價、三大法人、融資融券資料，使用方式如下：

下載台股基本資料

```
pipenv run python upload_data2mysql.py taiwan_stock_info
```

下載台股股價

```
pipenv run python upload_data2mysql.py taiwan_stock_price
```

下載台股三大法人

```
pipenv run python upload_data2mysql.py taiwan_stock_institutional_
investors
```

下載台股融資融券

```
pipenv run python upload_data2mysql.py taiwan_stock_margin_purchase_
short_sale
```

下載台股股東持股分級表

```
pipenv run python upload_data2mysql.py taiwan_stock_holding_shares_per
```

以上指令會下載資料，這需要一段時間，完成後，資料庫會如下圖：

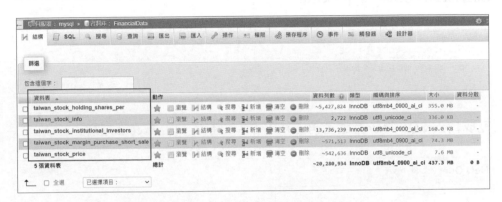

這時資料庫都有資料後，下一章節，開始製作第一個圖表。

11.6 製作第一個圖表

首先，到 Redash 的頁面，根據下圖操作，建立新的 Query：

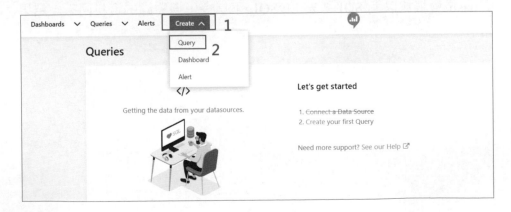

要如何建立新的 Query 呢？請讀者在以下圖表 1，輸入以下 SQL，撈出 2330 的收盤價與日期：

https://github.com/FinMind/FinMindBook/blob/master/DataEngineering/Chapter11/11.6/stock_price.sql

```
SELECT Close, Date
FROM taiwan_stock_price
WHERE StockID = '2330'
order by Date
```

完成輸入 SQL 後，請讀者再並按下 Save 儲存，與 Execute 執行：

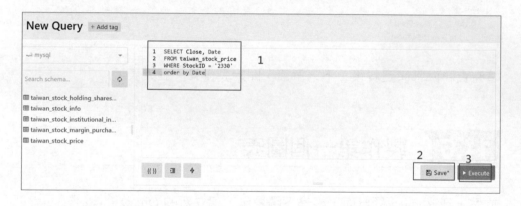

Redash 會根據以上 SQL，去 MySQL 下指令，撈對應的資料回來，結果如下：

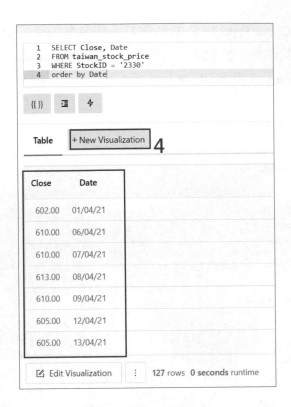

一開始只會回傳 table，如果想要做視覺化，請點選 4，New Visualization。

會看到上圖，Chart Type 代表圖形種類，因為是股價，選擇 Line，畫折線圖，而一般折線圖，X 軸會是 Date，在 X Column 選擇 Date，Y 軸是股價，在 Y Columns 選擇 Close：

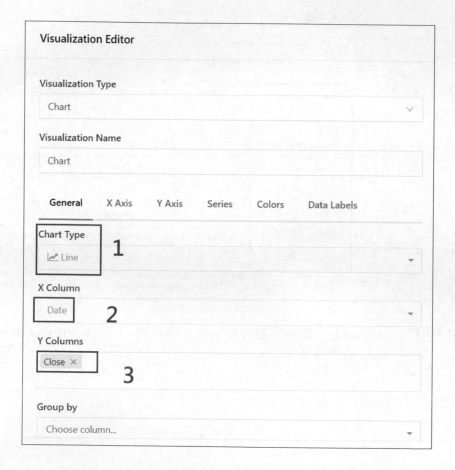

以下就是 Redash 畫出的折線圖，再點選 Save 儲存這張視覺化圖表：

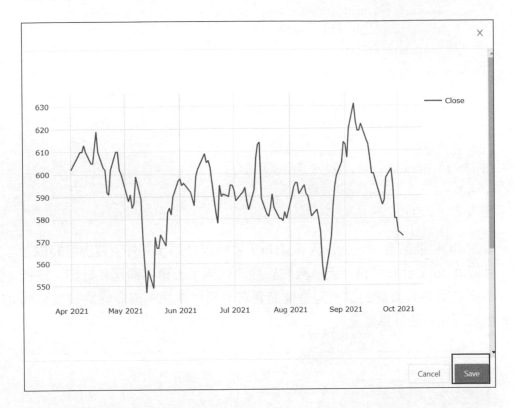

是不是很簡單？當讀者收集完資料後，後續只需要寫簡單的 SQL，就能做出視覺化圖表。但只看股價太單純了，進階一點的應用，想知道其他資料與股價的相關性，下一個圖表，本書串接股價與外資買賣的圖，觀察外資是否與股價有相關性。

步驟與創建股價 Query 相同，只需更改 SQL，SQL 如下：

https://github.com/FinMind/FinMindBook/blob/master/DataEngineering/
Chapter11/11.6/stock_price_vs_foreign_investor.sql

```
SELECT if(tsii.buy-tsii.sell>0, tsii.buy-tsii.sell,0) AS '買超',
       if(tsii.sell-tsii.buy>0, tsii.sell-tsii.buy,0) AS '賣超',
       stock_price.Close AS '股價',
       tsii.date AS date
FROM taiwan_stock_institutional_investors AS tsii
INNER JOIN taiwan_stock_info AS si ON si.stock_id = tsii.stock_id
INNER JOIN taiwan_stock_price AS stock_price ON tsii.stock_id =
stock_price.StockID
AND tsii.date = stock_price.Date
WHERE tsii.name = 'Foreign_Investor'
  AND si.stock_name = '台積電'
  AND tsii.date >= '2015-01-01'
```

以上 SQL 將股價與三大法人做 **Join**，才能同時顯示，**外資買賣與股價**，且同時 Join 台股資訊，因為相較於股票代碼，股票名稱比較好讀。這裡不單純看外資買賣的量，而是進而看買賣超的數據，因此對於買賣資料也進行額外的計算處理。

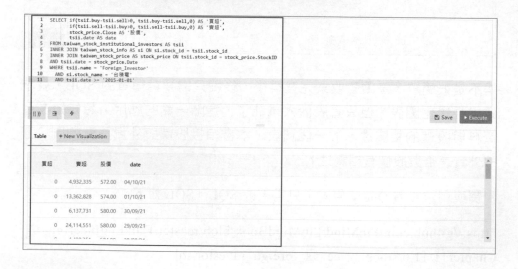

成功從 MySQL 撈資料後，接著點選 New Visualization，做圖：

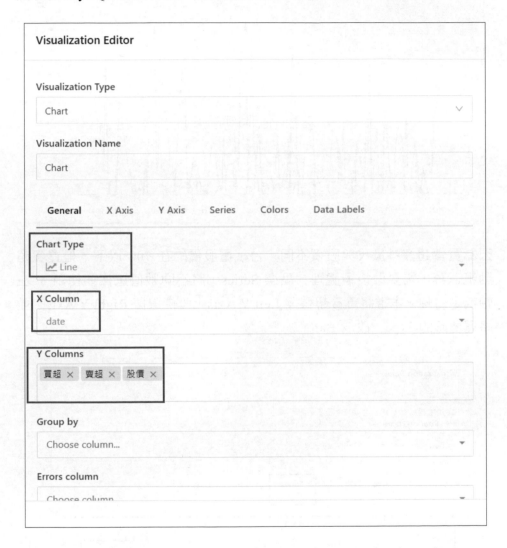

Chart Type 同樣選 Line，X Column 選 date，Y Columns 選擇買超、賣超、股價，呈現的圖表如下：

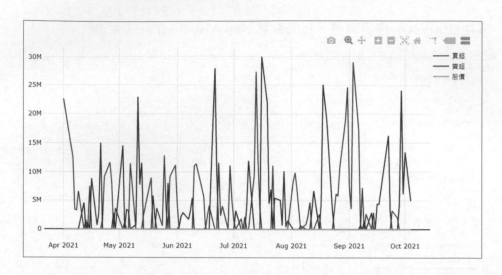

因為買賣超資料量尺與股價不同，台積電股價約在 700 以下，但買賣超動輒百萬，需要做後續處理。點選 Series，將不同數值選擇要使用 Y 左軸或是右軸，本書將買賣超選擇 Left Y Axis，股價選擇 Right Y Axis，並將買賣超的 Type，改成 Bar。

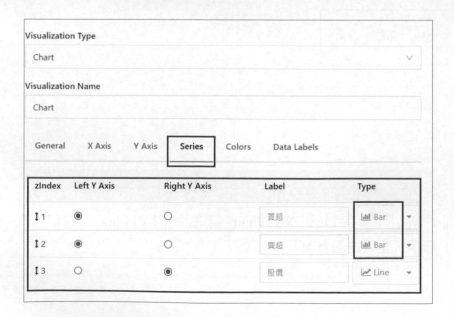

調整完後的圖表如下,相對來說,比較好閱讀。

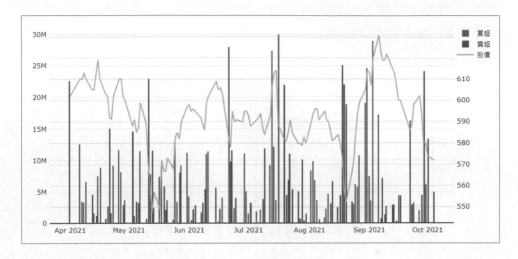

接著再調整顏色,一般而言,紅色代表買超、綠色代表賣超,設定如下,在 Colors 的部分去設定顏色,買超設定紅色,賣超設定綠色,股價就選擇藍色。

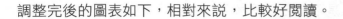

修改完後的圖表如下,是不是更好理解了,在股價大幅下降的期間,外
資大量賣超,相反,股價大幅上漲,剛好就是外資大量買超,最後點選
右下角的 Save,將此圖存檔。

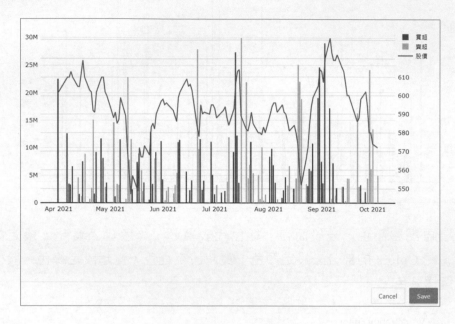

最後到左上角的位置,設定這個 Query 的名稱,本書設定是,股價 vs 外
資:

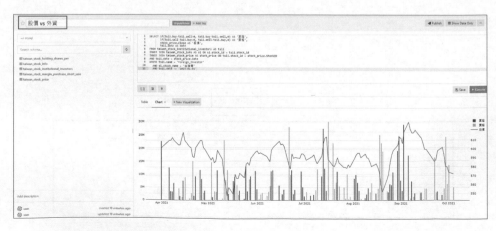

讀者可根據以上方法，額外增加，投信 vs 股價，SQL 如下，就不再帶著讀者一一設定了。

```
SELECT if(tsii.buy-tsii.sell>0, tsii.buy-tsii.sell,0) AS '買超',
       if(tsii.sell-tsii.buy>0, tsii.sell-tsii.buy,0) AS '賣超',
       stock_price.Close AS '股價',
       tsii.date AS date
FROM taiwan_stock_institutional_investors AS tsii
INNER JOIN taiwan_stock_info AS si ON si.stock_id = tsii.stock_id
INNER JOIN taiwan_stock_price AS stock_price ON tsii.stock_id = stock_
price.StockID
AND tsii.date = stock_price.Date
WHERE tsii.name = 'Investment_Trust'
  AND si.stock_name = '台積電'
  AND tsii.date >= '2015-01-02'
```

股價 vs 投信如下：

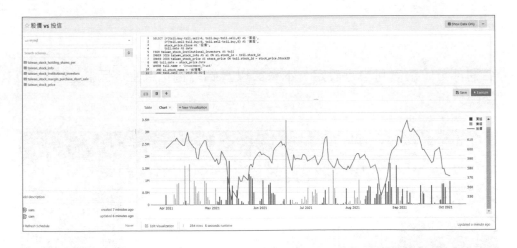

以上完成兩張圖表後，要如何在同一的畫面，看到多張圖表呢？下一章節，本書將介紹 dashboard，可以組合多種圖表。

11.7 第一個 Dashboard

前一章節，製作多個 Query，多種資料的視覺化，而 Dashboard，就是將多種 Query 的視覺化圖表，組合起來，方法如下，點選 Create，這次選擇 Dashboard。

名稱部分，讀者可自行設定，本章節以籌碼視覺化為主，因此這邊使用籌碼分析。

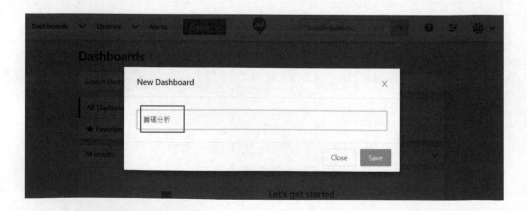

設定好名稱後，會進入以下畫面，點選右下角的 Add Widget，將前一章節的圖表加進來。

可以看到，以下三個，股價 vs 投信、股價 vs 外資、股價，剛好對應到前一章節設定的三個 Query，這裡選擇股價 vs 投信。

Choose Visualization 的意思是，選擇視覺化種類，分別有 Table 與 Chart，也對應到在 Query 開發中的 Table 與 Chart，本書在此選擇 Chart 圖表，之後點選 Add to Dashboard，加到 Dashboard 上。

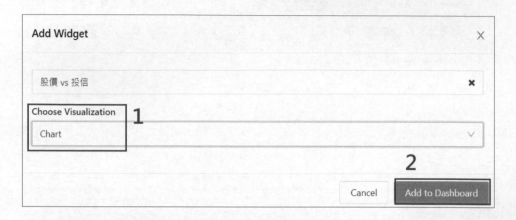

結果如下圖，成功將前一章節設定的 Query 圖表，加進 Dashboard 啦。

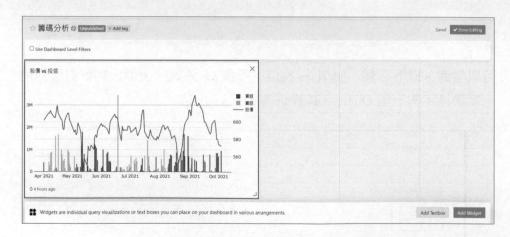

同樣的操作，將股價 vs 外資也加進來：

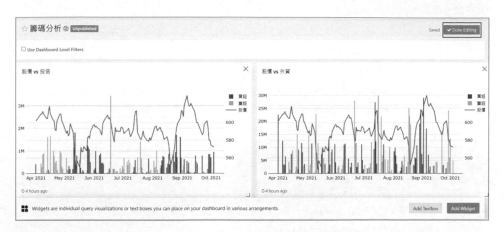

最後點選右上角，Done Editing，完成編輯，到此就完成第一個 Dashboard 了。

但目前為止，只能觀看台積電的分析，要如何輕易觀看其他股票呢？是否有一個地方，可以讓讀者直接選取想看得股票呢？以下筆者一一介紹。

11.8　設定下拉式選單

第一步，先建立股票清單，使用 Query，SQL 如下，因為筆者只操作非權證股票，因此以下 SQL 將權證類型的股票排除掉：

https://github.com/FinMind/FinMindBook/blob/master/DataEngineering/
Chapter11/11.6/stock_name_list.sql

```
SELECT DISTINCT
    `stock_name`
FROM
```

```
    `taiwan_stock_info`
WHERE
    (
        `industry_category` NOT IN(
            '認購權證(不含牛證)',
            '可展延牛證',
            '認售權證(不含熊證)',
            '熊證(不含可展延熊證)',
            '牛證(不含可展延牛證)',
            '不動產投資信託證券',
            '上櫃指數股票型基金(ETF)',
            '受益證券',
            '指數投資證券(ETN)',
            'ETN',
            '存託憑證',
            '封閉式基金'
        )
    )
```

將左上角的 Query 名稱，設為 stock name list，並點選右上角圖 3 的
Publish，將此 Query 設為公開，如果不設公開，後續無法使用。

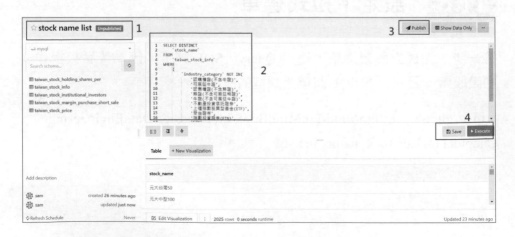

需要注意一點，筆者在此建立的是**股票名稱**清單，而非**股票代碼**，因為一般人相較於股票代碼，股票名稱比較好理解。

完成後，到股價 vs 外資的 Query 頁面，在下圖 1 的地方，將原本台積電的位置，改成，{{ 股票名稱 }}，這時 Redash 在圖 2 的地方，突然出現相對應的股票名稱，這有什麼用途呢？

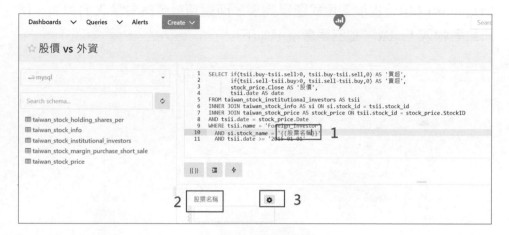

新長出來的部分，本書將會用此來設定下拉式選單，請讀者先點選 3 的地方，一個齒輪。

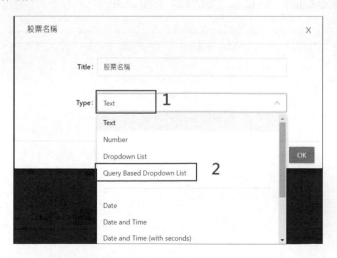

點選後,會跳出一個視窗,接著在點上圖 1,Text,會出現好幾個選項,
分別代表什麼意思呢?

Text:可以輸入任意文字。

Number:顧名思義,輸入數字。

Dropdown List:提供下拉式選單列表,而選項就是讀者輸入的列表,例
如讀者輸入,a,b,c,這三個選項,如下:

之後回到原先 Query 的畫面,就變成下拉式選單了。

```
1   SELECT if(tsii.buy-tsii.sell>0, tsii.buy-tsii.sell,0) AS '買超',
2       if(tsii.sell-tsii.buy>0, tsii.sell-tsii.buy,0) AS '賣超',
3       stock_price.Close AS '股價',
4       tsii.date AS date
5   FROM taiwan_stock_institutional_investors AS tsii
6   INNER JOIN taiwan_stock_info AS si ON si.stock_id = tsii.stock_id
7   INNER JOIN taiwan_stock_price AS stock_price ON tsii.stock_id = stock_price.StockID
8   AND tsii.date = stock_price.Date
9   WHERE tsii.name = 'Foreign_Investor'
10      AND si.stock_name = '{{股票名稱}}'
11      AND tsii.date >= '2015-01-01'
```

Query Based Dropdown List：顧名思義，根據 Query 的結果，產生的下拉式選單，與前一個 Dropdown List 相比，不需要一個個輸入選項，Redash 會直接讀取 Query 的結果，作為選項。

Date：日期欄位。

Date and Time：日期與時間欄位。

Date Range：日期區間欄位。

Date and Time Range：日期與時間區間欄位。

此章節需要下拉式選單，因此選擇 Query Based Dropdown List，之後在 Query 的地方，選擇 stock name list，最後點選 OK。

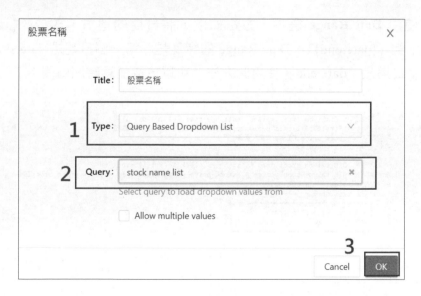

設定完成後，這時股票名稱，就成功變成下拉式選單了。

設定完股票名稱的下拉式選單後，如果想看不同時間區間，怎麼設定呢？

需要使用 **Date Range** 選項，方法如下，將日期的地方，改成，{{date.start}} 與 {{date.end}}，Date Range 的特性是，第一個日期是 date.start，第二個日期是，date.end，使用此方法，可以拉取任意時間區間：

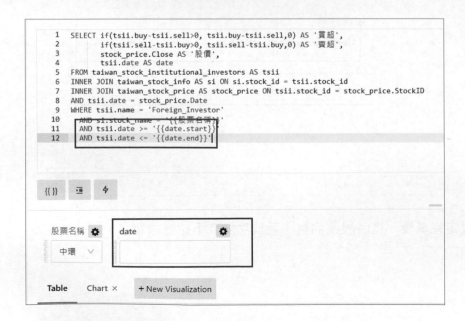

同樣的方式，點選 date 右上角的齒輪後，在 Type 的地方選取 Date Range。

這時就可以開始選取時間區間了。

本章節到此，已經將股價 vs 外資部分的 Query，完成下拉式選單，與時間區間的設定，麻煩讀者照著相同步驟，將股價 vs 投信的 Query 也做同樣的設定，可參考以下 SQL：

https://github.com/FinMind/FinMindBook/blob/master/DataEngineering/Chapter11/11.6/stock_price_vs_investment_trust_with_params.sql

```
SELECT if(tsii.buy-tsii.sell>0, tsii.buy-tsii.sell,0) AS '買超',
       if(tsii.sell-tsii.buy>0, tsii.sell-tsii.buy,0) AS '賣超',
       stock_price.Close AS '股價',
       tsii.date AS date
FROM taiwan_stock_institutional_investors AS tsii
INNER JOIN taiwan_stock_info AS si ON si.stock_id = tsii.stock_id
INNER JOIN taiwan_stock_price AS stock_price ON tsii.stock_id =
stock_price.StockID
AND tsii.date = stock_price.Date
WHERE tsii.name = 'Investment_Trust'
  AND si.stock_name = '{{股票名稱}}'
  AND tsii.date >= '{{date.start}}'
  AND tsii.date <= '{{date.end}}'
```

完成 Query 的設定後，開始設定 Dashboard 上的下拉式選單、日期選單，方法如下：

先跳轉到 Dashboard，籌碼分析的頁面。

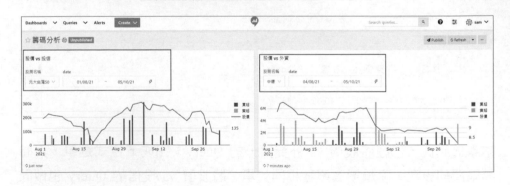

可以看到，兩張 Chart 都有相對應的下拉式選單、日期選單，但一個一個選取太笨了，有沒有辦法，設定 Dashboard 的參數呢？

先到股票 vs 投信的地方，點選右上角，三個點的地方後，選取 Edit
Parameters。

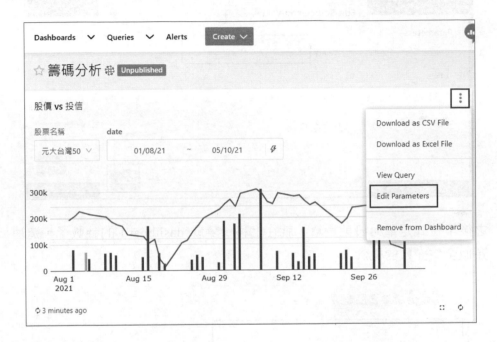

點選下圖的鉛筆。

將股票名稱與 date 的 Widget parameter，都改為 New dashboard parameter。

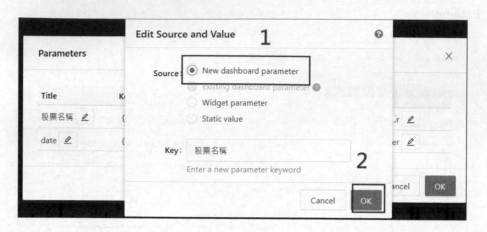

之後下拉式選單與日期選單，就往上移，變成 dashboard 的參數了，這裡類似全域變數的概念。

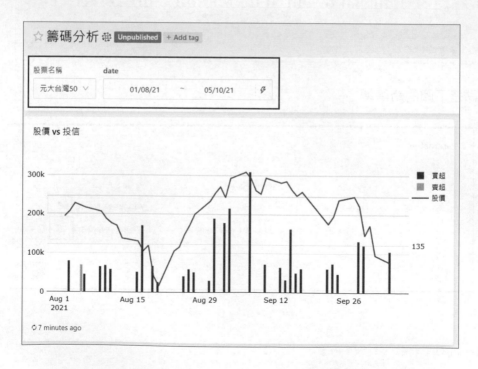

同樣的步驟,將股價 vs 外資的參數也做修改,這時要特別注意一點,前面使用的選項是 New dashboard parameter,新增一個 Dashboard 的參數,但在股價 vs 外資,就不能選擇 New dashboard parameter,因為此參數已經存在,因此選擇第二個,Existing dashboard parameter,使用現存的 Dashboard 參數。

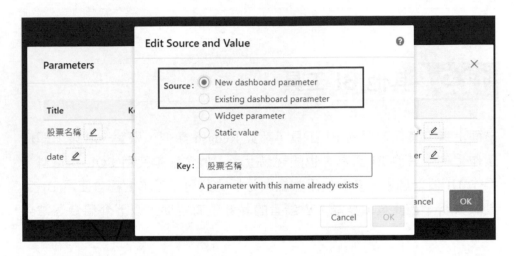

最後,兩個 Chart 的參數,都變成 Dashboard 的參數了。

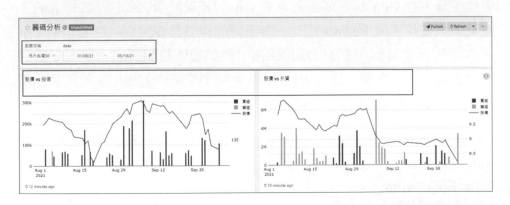

完成此步驟後,就可以任意對不同股票,切換圖表進行籌碼分析。

籌碼分析 Dashboard，目前只使用外資、投信，而在 11.5 章節中，本書下載了三大法人、融資融券、持股分級表，請讀者接續著將其他資料，都做到 Dashboard 上，整張 Dashboard 就會變得更加豐富，有助於後續分析。甚至照著本書方法，增加其他爬蟲，收集其他資料，將技術分析、籌碼分析、基本面分析，都做成不同的 Dashboard，製作專屬的看盤工具。

11.9　其他 BI 工具

市面上還有其他類似的 BI 工具，都與 Redash 類似，學會一套後，即使工作上使用其他 BI 工具，也能輕易上手，概念都是使用 SQL 撈資料，操作 UI 設計圖表，多數使用數據做分析的公司，基本上都會使用 BI 工具做分析，因此該項技能，對讀者的職涯很有幫助，以下介紹幾個常見的 BI 工具。

Superset：https://github.com/apache/superset，由 Apache 基金會開發的視覺化工具，在 Github 上有超過 40K stars，相較於 Redash 約 20K stars，這更多人使用。與 Redash 相比，有更好的權限控管機制。在公司中，不是每個人都能看到所有的 Dashboard，例如營收相關資訊，就會限定只有特定人士才能觀看，這時 BI 工具，就需要更細部的權限控管工具。因此，不同 BI 視覺化工具，根據讀者的需求選用即可。

Grafana：https://github.com/grafana/grafana，在 GitHub 上有超過 40 K stars，另一套視覺化工具，主要用於監控系統，製作監控儀表板，同樣可串接各種 ChatBot。

第 5 篇
排程管理工具

Chapter

12

排程管理工具 -
Apache Airflow

12.1 事前準備

12.1.1 Python 環境

由於 Airflow 快速發展，為了使用新版 Airflow，因此選擇 Python 3.7。但讀者不需太擔心，Python 3.6、3.7 在 pipenv 上並無太大差異。

使用 pyenv 安裝 Python 3.7 環境：

```
pyenv install miniconda3-4.7.12
```

在當前資料夾，設定使用 miniconda3-4.7.12 環境：

```
pyenv local miniconda3-4.7.12
```

完成以上設定後，接著執行 Python，就會顯示，使用 Python 3.7 啦！

```
sam@DESKTOP-IKT69L5:~/FinMindBook/DataEngineering/Chapter12/12.4$ python
Python 3.7.4 (default, Aug 13 2019, 20:35:49)
[GCC 7.3.0] :: Anaconda, Inc. on linux
Type "help", "copyright", "credits" or "license" for more information.
>>>
```

完成環境初始化後，開始介紹進入主題吧！

12.2　什麼是排程管理工具？

顧名思義，協助你管理排程的工具，那什麼是排程呢？

當你有任何一段程式碼、一個 Python 檔案，或是任何一個任務、工作，你想每天在固定時間執行，都能算是一種排程，這裡不限於 Python，任何程式都可以有 " 排程 "，而排程管理工具，有非常多種，如常見的：

- Windows 10 的工作排程器。
- Linux 內建的 crontab，算是一般工程師都會用的工具。
- Digdag，開源的排程管理系統，由 Java 撰寫，在 Github 上有 1.2k 個星星。
- Airflow，最知名的排程管理工具之一，由 Python 撰寫，在 Github 上有 29.2k 個星星。
- Prefect，相對新的排程管理工具，UI 介面非常具現代感，由 Python 撰寫，在 Github 上有 11.3k 個星星。

那有這麼多排程管理工具，初學者要如何做選擇呢？筆者的選擇是，Airflow，在下一個章節將會說明原因。

12.3 為什麼選擇 Airflow ？

筆者使用過 Crontab、Digdag、Airflow、Prefect，最後選擇 Airflow，有幾個原因。

1. 最知名、最多人使用。
2. 有良好的 UI 介面，能清楚知道目前流程進行的狀況。
3. 會根據 DAG 自動生成架構圖，對於大型 ETL、流程系統上，架構圖能讓人清楚的了解整個流程。
4. 良好的 Log 查看介面。
5. GCP、AWS 原生支援。

首先，第一點，可能也是最重要的一點，站在一個初階工程師的角度來看，可能不清楚這有什麼好處。但站在一個資深工程師，甚至管理者角度，團隊選擇的工具，必須讓團隊成員容易上手，甚至在尋找新成員時，知名的工具，能讓你更容易尋找到，擁有此項技能的夥伴。冷門的工具，既難維護，也找不到擁有此項技能的工程師。

而站在筆者的立場，介紹最知名的工具，能幫助到更多讀者，這也是我選擇的原因之一。

第二、三、四點，讓讀者在開發上，有良好的 UI 介面，更容易釐清整體架構，而站在產品開發上，架構圖絕對能讓新成員，更快速的融入產品開發上。

第五點就不用說了，兩大雲端廠商支援，讓開發者大大減少維運成本，不過本書依然會介紹，如何使用 Docker 做架設，即使大多數讀者，都沒有 GCP、AWS 環境，依然能學習如何使用 Airflow。

12.4　什麼是 Airflow ？

前面介紹完了，開始進入主題，Airflow 是一個排程管理工具，最大的特點是，DAG 的設計，在撰寫程式上，會製作工作流程的依賴關係，task a 執行完，再執行 task b、task c，最後執行 task d，如下，是一個基礎的 Airflow 流程範例：

task_a >> [task_b, task_c] >> task_d

最後 Airflow 會根據以上程式碼，製作出以下流程 DAG 圖。

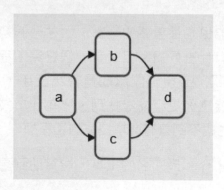

而 Airflow 最常被用於以下任務：

- ETL 的資料處理。
- 網路爬蟲、資料收集。

- 地端資料定時遷移到雲端。
- 資料邏輯運算。

可以發現，主要都是用於資料工程上，且近年大數據盛行，大大增加 Airflow 知名度。

接下來，下一個章節，將帶著讀者，架設第一個 Airflow，但在此之前，有幾個重點，先跟讀者說明。由於 Airflow 是由 Python 撰寫，因此筆者看過非常多人，使用 Airflow 的同時，也使用 Python 進行複雜運算、複雜的程式，將排程與任務混在一起，最後讓自己的程式變得非常複雜，難以維護。

排程，與要執行的工作、任務，必須有明確的分離、「職責隔離」，讓 Airflow 單純進行排程管理，其餘不論要做爬蟲、ETL、或是其他複雜的運算，都應該將程式碼隔離開，甚至運算資源也進行隔離。Airflow 就如同主管的腳色，負責管理各種工作，而自己本身，不直接參與工作執行。

12.5 架設第一個 Airflow

12.5.1 為什麼使用 Docker 架設？

架設 Airflow 有許多種方式，秉持著本書的理念，將使用 Docker 架設 Airflow。

可能會有讀者在這提出疑問，網路上很多教學，都是直接使用 Airflow 指令，建立 Airflow，為什麼本書要使用 Docker？本書到此章節，已經分享了許多 Docker 的好處，不過在此，還是說明使用 Docker 架設 Airflow 的好處。

- 容易部屬
  - 由於所需的服務、設定，全部使用 yml 檔，且容器化技術，能快速、「一致性」的部屬在不同環境。如開發環境（Dev）、測試環境（Stage）、前置生產環境（Pre-Prod）、生產環境（Prod）。
  - 私底下玩玩 Airflow，可能感受不到，但在公司、業界上，一定會有以上這些環境，如果你的環境不一致，很容易發生，開發環境上正常的程式，換版到產品上就出問題了。
- 高拓展性
  - 不論是本書使用的 Docker Swarm 或是常用的 K8S，對於 Docker，都能輕易進行拓展，如果只用 Airflow 指令在本地端架設，無法達到產品上，高效能的要求。
- 容易管理
  - 就如同本書前幾個章節，使用 Docker Swarm 統一管理爬蟲、API、資料庫、Redash 等，Airflow 使用 Docker，讓管理上更方便。

12.5.2　開始架設 Airflow

由於以上幾點，本書將使用 Docker Swarm 架設 Airflow，只需按照以下步驟，即可成功建立。

（在進行本節操作時，請先完成 7.5，建立 Docker Swarm、Portainer，並完成 7.6.1 的 MySQL 部屬。）

Airflow 需要資料庫才能使用，而本書 5.3 章節，使用的資料庫是 MySQL，因此在此同樣使用 MySQL，做為 Airflow 的資料庫。

1. 進入 MySQL，建立 airflow 資料庫。

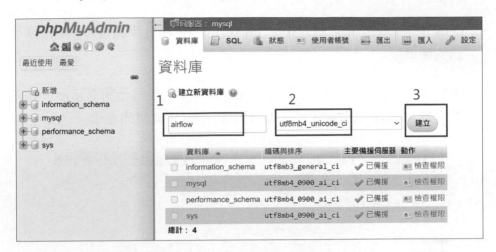

2. 幫 Airflow 建立 MySQL 的使用者帳號。請讀者根據以下步驟建立：

   步驟 1，切換到使用者帳號頁面。

   步驟 2，點選 " 新增使用者帳號 "。

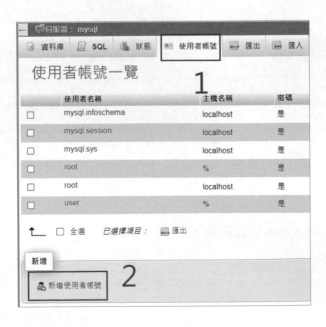

步驟 3，使用者名稱部分，填入 airflow。

步驟 4，本書密碼設定成 your_password，後續會教讀者如何設定成自己的密碼。

步驟 5，勾選，建立與使用者同名的資料庫，並授予所有權限，這是為了讓 airflow 這個使用者，擁有對 airflow 資料庫的權限。

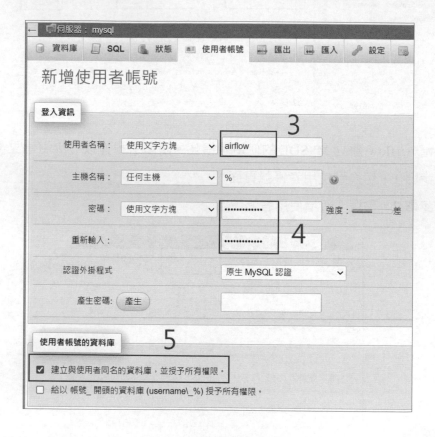

最後，步驟 6，滑到頁面最下面，點選執行。

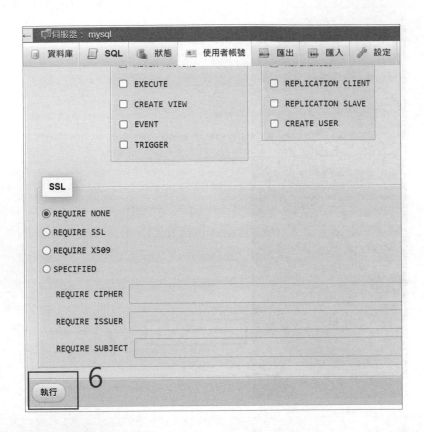

這時就完成權限設置了，接著使用 yml 檔建立 Airflow。

3. 使用以下連結下載 airflow.yml

https://github.com/FinMind/FinMindBook/blob/master/DataEngineering/
Chapter12/12.5/airflow.yml

（在 12.5.3 小節，會一一介紹 yml 中各項 services。）

在部屬 Airflow 前，先選定一台電腦 or 雲端設備，設定 node label，後續
將在這台電腦上，架設 Airflow，步驟如下：

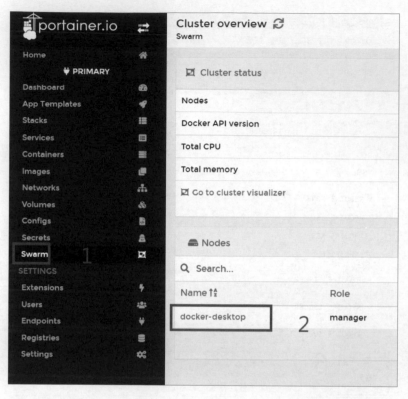

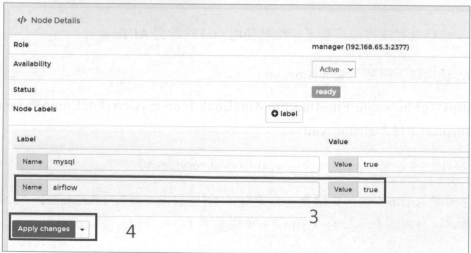

完成 Label 設定後，使用以下指令，建立 airflow。

```
docker stack deploy -c airflow.yml airflow
```

成功後，切換到 portainer 環境，正在初始化中。

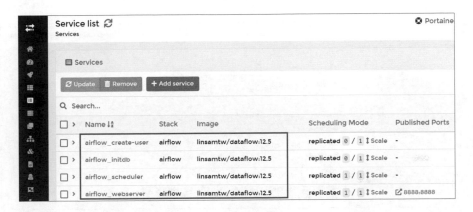

（airflow_create-user、airflow_initdb，執行完成後，replicated 會歸 0，這點讀者不用擔心，是正常現象。）

完成以上步驟後，先到 MySQL 的 airflow 資料庫查看，在初始化 Airflow 時，會在資料庫中建立相關 table。

接著跳轉到 Airflow 頁面，筆者先在自己的電腦，架設 Airflow，因此使用以下網址進入：

http://localhost:8888/

進入後，會看到以下畫面，帳號密碼是：admin／admin

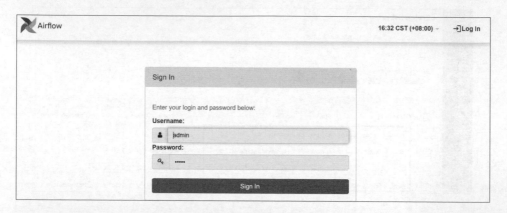

登入後，在 DAG 底下，會看到一個 HelloWorld 的 DAG，這是筆者提供的範例。

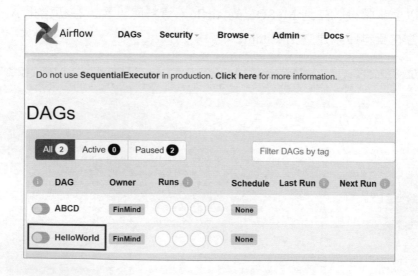

點擊 HelloWorld 後，會進到 DAG 畫面，按照以下操作，來執行第一個
DAG。

步驟 1，先把開關打開，接著步驟 2、3，執行 HelloWorld DAG。

完成後，注意看下圖，會產生一個正在執行的任務，不同顏色代表不同
的狀態：

- 先從灰色 - queued，等待中。
- 變成亮綠色 running，執行中。
- 完成任務後，變成深綠色 success。

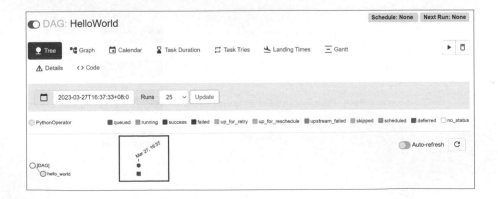

滑鼠滑到下面的正方形，可以看到當下狀態。

點擊後，會跳出以下視窗，可以查看 hello_world 這個任務的各種資訊，如 Details、Log，在此筆者先點擊 Log 進行查看。

Log 如下：

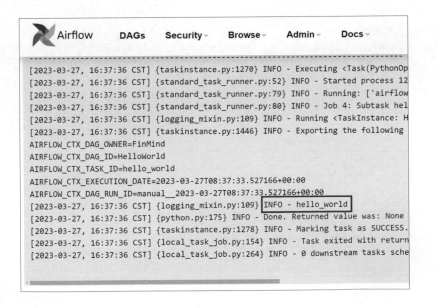

放大來看，可以看到：

在此顯示的 hello_world，就是筆者在此任務進行的 Python 程式，如下連結：

https://github.com/FinMind/FinMindBook/blob/master/DataEngineering/Chapter12/12.5/dataflow/etl/hello_world.py

（程式細節，在下個章節，將會詳細介紹。）

在此就完成架設 Airflow，並成功執行 DAG 了，接著筆者將介紹本書 Airflow 的架構。

12.5.3　架構介紹

以下連結是 12.5 的架構：

https://github.com/FinMind/FinMindBook/tree/master/DataEngineering/Chapter12/12.5

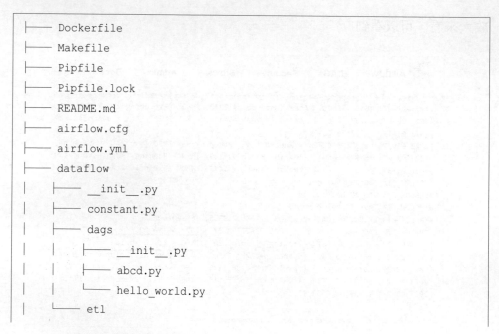

```
├── Dockerfile
├── Makefile
├── Pipfile
├── Pipfile.lock
├── README.md
├── airflow.cfg
├── airflow.yml
├── dataflow
│   ├── __init__.py
│   ├── constant.py
│   ├── dags
│   │   ├── __init__.py
│   │   ├── abcd.py
│   │   └── hello_world.py
│   └── etl
```

```
|            ├──── __init__.py
|            ├──── abcd.py
|            └──── hello_world.py
├──── mysql.yml
├──── portainer.yml
└──── setup.py
```

架構基本上與前面幾個章節一樣，都有最基本的 Dockerfile、Makefile、
Pipfile 等等，本書將一一做介紹。

❑ Pipfile

https://github.com/FinMind/FinMindBook/blob/master/DataEngineering/
Chapter12/12.5/Pipfile

```
[[source]]
url = "https://pypi.org/simple"
verify_ssl = true
name = "pypi"

[packages]
apache-airflow = "==2.2.5"
pymysql = "==1.0.2"
dataflow = {editable = true, path = "."}

[dev-packages]

[requires]
python_version = "3.7"
```

本章節使用的 Pipfile，與前幾個章節不同。如 12.0 所說，在此使用
Python 3.7 的版本，Pipfile 好處是，連 Python 版本都會自動作紀錄。

❏ Dockerfile

https://github.com/FinMind/FinMindBook/blob/master/DataEngineering/
Chapter12/12.5/Dockerfile

```
# 由於 continuumio/miniconda3:4.3.27 中的 Debian
# 版本太舊，因此改用 ubuntu 系統
FROM ubuntu:18.04

# 系統升級、安裝 python
RUN apt-get update && apt-get install python3.7 -y && apt-get install
python3-pip -y

RUN mkdir /FinMindProject
COPY . /FinMindProject/
COPY ./airflow.cfg /FinMindProject/dataflow/airflow.cfg
WORKDIR /FinMindProject/

# env
ENV LC_ALL=C.UTF-8
ENV LANG=C.UTF-8

# install package
RUN pip3 install pipenv==2020.6.2
RUN pipenv sync

# airflow
ARG AIRFLOW_USER_HOME=/FinMindProject/dataflow
ARG AIRFLOW_DEPS=""
ARG PYTHON_DEPS=""
ENV AIRFLOW_HOME=${AIRFLOW_USER_HOME}

WORKDIR /FinMindProject/
```

由於本章節使用 Python 3.7，因此安裝 Python 3.7，其餘設定與前面章節都相同，相信讀者看到此，應該非常熟悉 Dockerfile 了。

但在此有一個不同的地方：

```
COPY ./airflow.cfg /FinMindProject/dataflow/airflow.cfg
```

在建立 Airflow 時，需要設定檔，因此在建立 Docker Image 時，將相關設定檔，複製進去，那設定檔中有什麼資訊呢？筆者將一一介紹。

❏ airflow.cfg

airflow.cfg 設定檔超過 200 個參數可做調整，但讀者不用擔心，這麼多參數會讀不完，初期，先不用了解所有參數，有需求再參考官方文件即可。

https://airflow.apache.org/docs/apache-airflow/2.2.5/configurations-ref.html

本書中，整理出幾個常用的參數：

1. core 層

https://github.com/FinMind/FinMindBook/blob/master/DataEngineering/
Chapter12/12.5/airflow.cfg#L1

```
[core]
dags_folder = /FinMindProject/dataflow/dags
hostname_callable = socket.getfqdn
default_timezone = Asia/Taipei
executor = SequentialExecutor
sql_alchemy_conn = mysql+pymysql://airflow:your_password@mysql/airflow
sql_engine_encoding = utf-8
sql_alchemy_pool_enabled = True
sql_alchemy_pool_size = 5
sql_alchemy_max_overflow = 10
```

```
sql_alchemy_pool_recycle = 1800
sql_alchemy_pool_pre_ping = True
sql_alchemy_schema =
parallelism = 32
max_active_tasks_per_dag = 16
dags_are_paused_at_creation = True
max_active_runs_per_dag = 16
```

a. dags_folder = /FinMindProject/dataflow/dags

dag 的程式碼路徑。

而本書的 dag 是放在以下路徑，dataflow/dags（如架構圖），最後在 Dockerfile 將程式碼複製到 /FinMindProject 路徑中，因此在此 dags_folder 設定成 /FinMindProject/dataflow/dags。讀者如果開發自己的專案，可自行調整 dags 的位置。

b. sql_alchemy_conn = mysql+pymysql://airflow:your_password@mysql/airflow

資料庫位置。

還記得嗎？在本書 12.5.2 章節中，建立 airflow 資料庫，並新增使用者帳號，那要如何告訴 Airflow 呢？靠的就是 sql_alchemy_conn 的設定。讀者後續可自行調整 airflow 的帳號、密碼。

c. default_timezone = Asia/Taipei

時區設定。

本書中設定成台北時區。

d. executor = SequentialExecutor

設定 Airflow 要使用哪個執行引擎。可根據不同情境做選擇，本章節使用 SequentialExecutor 為例，其餘可選的引擎有以下幾種：

i. SequentialExecutor：單機執行，並無分散式功能，適合初期使用。

ii. LocalExecutor：一樣是單機執行，但可支援多執行緒。

iii. CeleryExecutor：如本書中 5.5.3 介紹的 Celery 分散式，使用分散式執行任務。

iv. DaskExecutor：另一套分散式工具，主要用於大數據分析（相對上，celery 是用於執行一般任務）。

v. KubernetesExecutor：基於 Kubernetes 的原生 Airflow 分散式架構。

vi. CeleryKubernetesExecutor：Kubernetes 結合 Celery 的分散式架構（相較於 KubernetesExecutor，結合 Celery 後拓展性更高，但也相對複雜）。

e. max_active_tasks_per_dag = 16

每個 DAG 中，最多同時執行 task 的數量。

在某些情境下（如雲端機器效能不足），你不希望同時執行太多 task，則可以調整此設定。

f. max_active_runs_per_dag = 16

最多同時執行 DAG 的數量。

當專案慢慢長大後，例如超過 100 個 DAG，這時可能因為超過雲端機器負荷上限，需要限制同一時間，執行 DAG 的數量。

2. celery 層

還記得本書 5.5.3 中的 celery 嗎？在 Airflow 也是同一套概念，如果在 core 層中，設定 executor = CeleryExecutor，則需要對 celery 做進一步設定（在此由於使用 SequentialExecutor，因此不會使用 celery 的設定，到 12.8 會真實展示一個 CeleryExecutor 的例子）。

https://github.com/FinMind/FinMindBook/blob/master/DataEngineering/Chapter12/12.5/airflow.cfg#L196

```
[celery]
celery_app_name = airflow.executors.celery_executor
worker_concurrency = 16
worker_umask = 0o077
broker_url = redis://redis:6379/0
result_backend = redis://redis:6379/1
```

a. worker_concurrency = 16

最多可以有 16 個 worker 工人同時工作。這取決於整個分散式架構中 celery 的數量（如 7.6.3 只設定 1 個 celery_twse，當然就只能有一個工人同時工作，如需多個工人，只需調整 7.6.3 crawler.yml 檔中的 replicas 數量即可）。

b. broker_url = redis://redis:6379/0

還記得本書 5.5.2 的 RabbitMQ 嗎？ RabbitMQ 扮演 Broker 腳色，而在 Airflow 中，選用 redis 扮演 Broker 腳色，讀者這時可能會有疑問，什麼是 redis，筆者將在 Chapter 13 做介紹。

c. result_backend = redis://redis:6379/1

任務運行完後的結果，需要有一個地方做儲存。在 Airflow 架構下，不同 task 中的結果，可以互相傳遞，如下圖。

任務 a 可以將執行完的結果，傳遞給任務 b、c、d，這裡 a,b,c,d 泛指任何任務，這讓開發上，有非常大的彈性。

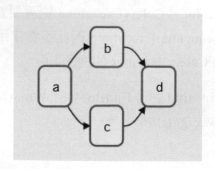

以上 core、celery 層，算是初期最常使用的參數，也是最容易需要做調整
的。接著介紹，airflow.yml。

❑ airflow.yml

https://github.com/FinMind/FinMindBook/blob/master/DataEngineering/
Chapter12/12.5/airflow.yml

```yaml
version: '3.0'
services:
  initdb:
    image: linsamtw/dataflow:12.5
    command: pipenv run airflow db init
    restart: on-failure
    # swarm 設定
    deploy:
      mode: replicated
      replicas: 1
      placement:
        constraints: [node.labels.airflow == true]
    networks:
      - my_network

  create-user:
    image: linsamtw/dataflow:12.5
    command: pipenv run airflow users create --username admin --firstname
lin --lastname sam --role Admin -p admin --email finmind.tw@gmail.com
    depends_on:
      - initdb
    restart: on-failure
    # swarm 設定
    deploy:
```

```
    mode: replicated
    replicas: 1
    placement:
      constraints: [node.labels.airflow == true]
  networks:
    - my_network

webserver:
  image: linsamtw/dataflow:12.5
  hostname: "airflow-webserver"
  command: pipenv run airflow webserver -p 8888
  depends_on:
    - initdb
  restart: always
  ports:
    - 8888:8888
  environment:
    - TZ=Asia/Taipei
# swarm 設定
  deploy:
    mode: replicated
    replicas: 1
    placement:
      constraints: [node.labels.airflow == true]
  networks:
    - my_network

scheduler:
  image: linsamtw/dataflow:12.5
  hostname: "airflow-scheduler"
  command: pipenv run airflow scheduler
  depends_on:
```

```
    - initdb
  restart: always
  # swarm 設定
  deploy:
    mode: replicated
    replicas: 1
    placement:
      constraints: [node.labels.airflow == true]
  environment:
    - TZ=Asia/Taipei
  networks:
      - my_network

networks:
  my_network:
    external: true
```

■ initdb，初始化資料庫。

```
command: pipenv run airflow db init
```

還記得在 12.5.2 中，MySQL 的 airflow 資料庫中，新增了許多 table 嗎？
就是這個 services 建立的。

■ create-user，建立使用者帳號。

```
command: pipenv run airflow users create --username admin --firstname lin
--lastname sam --role Admin -p admin --email
```

還記得在 12.5.2 中，使用 admin / admin 登入 airflow 嗎？就是在此做的
設定，讀者可自行調整使用者名稱、密碼等資訊。

■ webserver

```
command: pipenv run airflow webserver -p 8888
```

Airflow 的 UI 介面，方便開發者，查看各個任務運行狀況，本書中使用 8888 port，讀者可自行調整。

■ scheduler

```
command: pipenv run airflow scheduler
```

Airflow 是排程管理系統，因此 scheduler 就是負責管理排程。

以上，建立完 Airflow 了，也介紹了 Airflow 各項設定、docker 服務，那要如何撰寫一個 DAG 呢？筆者將帶讀者進入，dags 資料夾內，探索第一個範例 - hello_world.py。

在介紹 code 之前，先介紹該專案的架構，主要由 dags、etl 兩個資料夾組成。

```
├──── dataflow
│     ├──── __init__.py
│     ├──── constant.py
│     ├──── dags
│     │     ├──── __init__.py
│     │     ├──── abcd.py
│     │     └──── hello_world.py
│     └──── etl
│           ├──── __init__.py
│           ├──── abcd.py
│           └──── hello_world.py
```

■ dags 資料夾中，負責建立 dag，並「管理」任務。
■ etl 資料夾中，負責「實作」任務（Airflow 大多用於 ETL 資料處理，因此在此，資料夾命名為 etl）。

秉持著本書的原則，職責分離，在此將「管理」與「實作」拆開。

以本章節範例，hello_world 為例。

❑ etl/hello_world.py

https://github.com/FinMind/FinMindBook/blob/master/DataEngineering/
Chapter12/12.5/dataflow/etl/hello_world.py

```
from airflow.operators.python_operator import (
    PythonOperator,
)


def create_hello_world_task() -> PythonOperator:
    return PythonOperator(
        task_id="hello_world",
        python_callable=lambda: print(
            "hello_world"
        ),
    )
```

在此的任務是，印出 hello_world 字串，因此實作的邏輯，如上程式碼。

（如果讀者對於 PythonOperator 有疑問，將在下個章節介紹，基本上就是
用於建立 Python 任務的函數。）

❑ dags/hello_world.py

https://github.com/FinMind/FinMindBook/blob/master/DataEngineering/
Chapter12/12.5/dataflow/dags/hello_world.py

```
import airflow

from dataflow.constant import (
```

```
    DEFAULT_ARGS,
    MAX_ACTIVE_RUNS,
)
from dataflow.etl.hello_world import (
    create_hello_world_task,
)

with airflow.DAG(
    dag_id="HelloWorld",
    default_args=DEFAULT_ARGS,
    schedule_interval=None,
    max_active_runs=MAX_ACTIVE_RUNS,
    catchup=False,
) as dag:
    create_hello_world_task()
```

https://github.com/FinMind/FinMindBook/blob/master/DataEngineering/
Chapter12/12.5/dataflow/constant.py

```
import datetime

DEFAULT_ARGS = {
    "owner": "FinMind",
    "retries": 1,
    "retry_delay": datetime.timedelta(
        minutes=1
    ),
    "start_date": datetime.datetime(
        2022, 1, 1
    ),
    "execution_timeout": datetime.timedelta(
        minutes=60
    ),
```

```
    "max_active_tasks": 10
}
MAX_ACTIVE_RUNS = 1
```

在 etl 資料內，實作完任務後，再到 dag 資料夾內，引入此任務，而 dag
所需參數，由於變動性不高，統一放到 constant.py（own、retries 次數等
設定，基本上不需客製化）。

此架構好處是，在 etl 層，可以根據不同需求，建立不同任務，最後再由
dag 層統一管理，甚至設定相依性。

再介紹另一個例子，在 etl 層，分別建立 a、b、c、d 四個任務，最後再
由 dag 層，管理流程，如下：

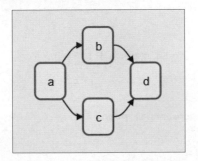

https://github.com/FinMind/FinMindBook/blob/master/DataEngineering/
Chapter12/12.5/dataflow/etl/abcd.py

```
from airflow.operators.python_operator import (
    PythonOperator,
)



def create_a_task() -> PythonOperator:
    return PythonOperator(
```

```
        task_id="a",
        python_callable=lambda: print(
            "a"
        ),
    )


def create_b_task() -> PythonOperator:
    return PythonOperator(
        task_id="b",
        python_callable=lambda: print(
            "b"
        ),
    )


def create_c_task() -> PythonOperator:
    return PythonOperator(
        task_id="c",
        python_callable=lambda: print(
            "c"
        ),
    )


def create_d_task() -> PythonOperator:
    return PythonOperator(
        task_id="d",
        python_callable=lambda: print(
            "d"
        ),
    )
```

https://github.com/FinMind/FinMindBook/blob/master/DataEngineering/
Chapter12/12.5/dataflow/dags/abcd.py

```python
import airflow

from dataflow.constant import (
    DEFAULT_ARGS,
    MAX_ACTIVE_RUNS,
)
from dataflow.etl.abcd import (
    create_a_task,
    create_b_task,
    create_c_task,
    create_d_task,
)

with airflow.DAG(
    dag_id="ABCD",
    default_args=DEFAULT_ARGS,
    schedule_interval=None,
    max_active_runs=MAX_ACTIVE_RUNS,
    catchup=False,
) as dag:
    task_a = create_a_task()
    task_b = create_b_task()
    task_c = create_c_task()
    task_d = create_d_task()
    task_a >> [task_b, task_c] >> task_d
```

基本上 a、b、c、d 可以替代成，讀者任意的任務、程式碼。

以上建立了 Airflow，也介紹了 Airflow 環境設定檔 airflow.cfg，並建立
第一個 DAG - hello_world，下個小節，本書將介紹一些 Airflow 常用函
數的意義。

12.6 DAG 介紹

DAG 是 Airflow 中，最小的排程單元。不同的排程，使用不同的 DAG，管理執行時間，如每小時執行、每天執行等等。而 DAG 底下，由一系列的任務 task 組成。

以 12.5 的 abcd 為例。

https://github.com/FinMind/FinMindBook/blob/master/DataEngineering/Chapter12/12.5/dataflow/dags/abcd.py

```python
import airflow

from dataflow.constant import (
    DEFAULT_ARGS,
    MAX_ACTIVE_RUNS,
)
from dataflow.etl.abcd import (
    create_a_task,
    create_b_task,
    create_c_task,
    create_d_task,
)

with airflow.DAG(
    dag_id="ABCD",
    default_args=DEFAULT_ARGS,
    schedule_interval=None,
    max_active_runs=MAX_ACTIVE_RUNS,
    catchup=False,
) as dag:
```

```
task_a = create_a_task()
task_b = create_b_task()
task_c = create_c_task()
task_d = create_d_task()
task_a >> [task_b, task_c] >> task_d
```

以上是，建立一個 ABCD 的 dag 程式碼，筆者在此整理出幾個常用的參數。

- dag_id
  - DAG 的名稱，需注意一點，在整個 DAG 中，不能重複。
- default_args
  - 預設參數，常用的參數有：
    - ▶ owner - 該 DAG 所屬的使用者。
      - ❖ 假設一個 Airflow 系統中，包含爬蟲收集資料、ETL 資料清理、計算出各項指標、各種策略、最後使用 chatbot 發出交易訊號，則會在不同階段，設定不同 owner。
    - ▶ retries - 重試次數。
      - ❖ DAG 中的任務可能因不明原因失敗，則可使用此參數，讓該任務重新執行。
    - ▶ retry_delay - 重試任務時，需間隔多久。
    - ▶ start_date - DAG 開始時間。
      - ❖ 多數情境都是，建立 DAG 當下，立刻使用排程開始進行。
    - ▶ execution_timeout - DAG 中，任務的超時時間。
      - ❖ 當執行超過此設定，將會被強制停止，避免持續運行，占用其他任務的資源。
    - ▶ max_active_tasks - 最多可同時執行任務的數量。
- scheduler_interval

- 排程的執行頻率。可使用 crontab 相同設定格式，如 */5 * * * *，代表每 5 分鐘執行一次。
- max_active_runs
  - 設定最多可同時執行多少個 DAG。
- catchup
  - 當 DAG 建立時，是否要回朔到 start_date。
  - 例如 start_date 是 2020-01-01，但 DAG 是 2023-01-01 建立。如果此設定為 True，則會重新執行，2020-01-01~2023-01-01 所有任務。

以上就是 DAG 最常用的幾個參數，而該 DAG 要執行的任務，就是在 with 底下，建立的 task_a、task_b、task_c、task_d。

相信第一次接觸 Airflow 的讀者，都會有一個問題：

```
task_a >> [task_b, task_c] >> task_d
```

這是什麼意思？

這是在 Airflow 中，建立一個 DAG 時，特有的寫法，直接看圖會更好理解。

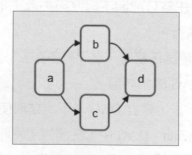

代表任務執行順序為，a 先開始，接著 b 跟 c 同時執行，等 b、c 執行完成後，最後再執行任務 d。

DAG 介紹完後，接著介紹，如何建立一個任務。

12.7 常見 Operator 介紹

在 Airflow 中，統一用 Operator 建立任務物件，Operator 的範圍非常廣泛，在 12.7 節，將介紹幾個最常被使用的 Operator。

在介紹之前，請讀者先按照以下步驟，更新 Airflow。

❏ 下載 airflow.yml

https://github.com/FinMind/FinMindBook/blob/master/DataEngineering/Chapter12/12.7/airflow.yml

```
version: '3.0'
services:
  initdb:
    image: linsamtw/dataflow:12.7
    command: pipenv run airflow db init
    restart: on-failure
    # swarm 設定
    deploy:
      mode: replicated
      replicas: 1
      placement:
        constraints: [node.labels.airflow == true]
    networks:
        - my_network

  create-user:
    image: linsamtw/dataflow:12.7
    command: pipenv run airflow users create --username admin --firstname
lin --lastname sam --role Admin -p admin --email finmind.tw@gmail.com
```

```
    depends_on:
      - initdb
    restart: on-failure
    # swarm 設定
    deploy:
      mode: replicated
      replicas: 1
      placement:
        constraints: [node.labels.airflow == true]
    networks:
        - my_network

webserver:
    image: linsamtw/dataflow:12.7
    hostname: "airflow-webserver"
    command: pipenv run airflow webserver -p 8888
    depends_on:
      - initdb
    restart: always
    ports:
        - 8888:8888
    environment:
      - TZ=Asia/Taipei
    # swarm 設定
    deploy:
      mode: replicated
      replicas: 1
      placement:
        constraints: [node.labels.airflow == true]
    networks:
        - my_network
```

```
  scheduler:
    image: linsamtw/dataflow:12.7
    hostname: "airflow-scheduler"
    command: pipenv run airflow scheduler
    depends_on:
      - initdb
    restart: always
    # swarm 設定
    deploy:
      mode: replicated
      replicas: 1
      placement:
        constraints: [node.labels.airflow == true]
    environment:
      - TZ=Asia/Taipei
    networks:
        - my_network

networks:
  my_network:
    external: true
```

更新 Airflow

```
docker stack deploy -c airflow.yml airflow
```

完成以上步驟後，切換到 Airflow 頁面：

http://localhost:8888

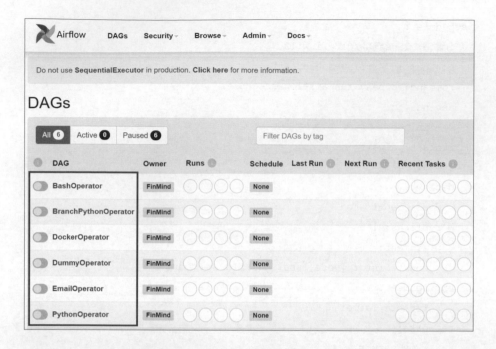

可以看到有 6 個 Operator，以下將介紹這 6 個。

12.7.1 BashOperator

用於建立一個，執行 Bash or Shell 指令的任務。例如某些 linux 指令、工具，Python 並無相關的 Package，則可以使用 BashOperator 執行。以下是本書範例。

❏ etl/bash_operator.py

https://github.com/FinMind/FinMindBook/blob/master/DataEngineering/Chapter12/12.7/dataflow/etl/bash_operator.py

```
from airflow.operators.bash_operator import (
    BashOperator,
```

```
)


def create_bash_operator_task() -> BashOperator:
    return BashOperator(
        task_id="BashOperator",
        bash_command="echo BashOperator",
    )
```

使用 bash_command 參數,執行 bash 指令。建立完任務後,接著建立 DAG。

❑ dags/bash_operator.py

https://github.com/FinMind/FinMindBook/blob/master/DataEngineering/ Chapter12/12.7/dataflow/dags/bash_operator.py

```
import airflow

from dataflow.constant import (
    DEFAULT_ARGS,
    MAX_ACTIVE_RUNS,
)
from dataflow.etl.bash_operator import (
    create_bash_operator_task,
)


with airflow.DAG(
    dag_id="BashOperator",
    default_args=DEFAULT_ARGS,
    schedule_interval=None,
    max_active_runs=MAX_ACTIVE_RUNS,
    catchup=False,
```

```
) as dag:
    create_bash_operator_task()
```

建立完 DAG 後，跳轉到 Airflow 頁面，根據以下步驟，實際操作一遍，
更好理解。

1. 進入 BashOperator 的 DAG 中。

2. 點擊按鈕，啟動此 DAG。

3. 執行。

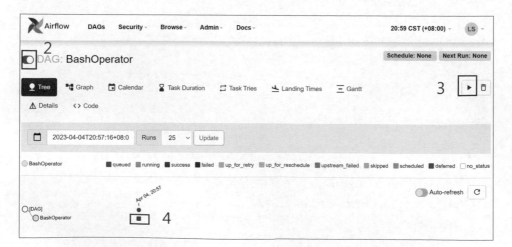

4. 執行完 DAG 後，等待約 10 秒，如果結果如圖上，代表執行成功。接著點選正方形。

5. 查看 Log。

6. 可以看到，圖 6 就是執行的指令。

7. 在 7 個位置，就是該指令輸出的結果。

以上就是 BashOperator，如何建立、且執行的情境，接著介紹 DummyOperator。

12.7.2 PythonOperator

用於建立一個，執行 Python 程式的任務，例如爬蟲任務。以下是一個建立 Python Operator 的範例。

❏ etl/python_operator.py

https://github.com/FinMind/FinMindBook/blob/master/DataEngineering/
Chapter12/12.7/dataflow/etl/python_operator.py

```
from airflow.operators.python_operator import (
    PythonOperator,
```

```
)


def crawler():
    print("crawler")



def create_python_operator_task() -> PythonOperator:
    return PythonOperator(
        task_id="PythonOperator",
        python_callable=crawler,
    )
```

在 python_callable 中，提供要執行的 python 函數即可，在此是印出 crawler 字串，讀者可自行修改成，爬蟲函數，或是直接跳到 12.8 章節，本書將結合爬蟲程式。

執行方式與 BashOperator 一樣：

1. 點擊按鈕打開。
2. 執行。
3. 等執行完後，查看 Log。

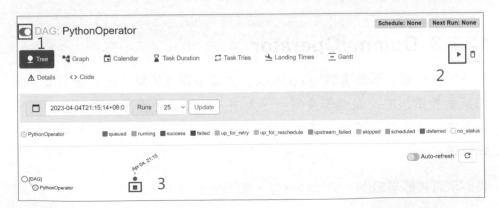

4. 最後印出 crawler 字串。

```
     Airflow        DAGs      Security ▾      Browse ▾      Admin ▾      Docs ▾

[2023-04-04, 21:21:22 CST] {taskinstance.py:1245} INFO
--------------------------------------------------------------------------------
[2023-04-04, 21:21:22 CST] {taskinstance.py:1250} INFO - Starting attempt 2 of 3
[2023-04-04, 21:21:22 CST] {taskinstance.py:1251} INFO -
--------------------------------------------------------------------------------
[2023-04-04, 21:21:22 CST] {taskinstance.py:1270} INFO - Executing <Task(PythonOp
[2023-04-04, 21:21:22 CST] {standard_task_runner.py:52} INFO - Started process 41
[2023-04-04, 21:21:22 CST] {standard_task_runner.py:79} INFO - Running: ['airflow
[2023-04-04, 21:21:22 CST] {standard_task_runner.py:80} INFO - Job 13: Subtask Py
[2023-04-04, 21:21:22 CST] {logging_mixin.py:109} INFO - Running <TaskInstance: P
[2023-04-04, 21:21:22 CST] {taskinstance.py:1448} INFO - Exporting the following
AIRFLOW_CTX_DAG_OWNER=FinMind
AIRFLOW_CTX_DAG_ID=PythonOperator
AIRFLOW_CTX_TASK_ID=PythonOperator
AIRFLOW_CTX_EXECUTION_DATE=2023-04-04T13:15:14.200366+00:00
AIRFLOW_CTX_DAG_RUN_ID=manual__2023-04-04T13:15:14.200366+00:00              4
[2023-04-04, 21:21:22 CST] {logging_mixin.py:109} INFO - crawler
[2023-04-04, 21:21:22 CST] {python.py:175} INFO - Done. Returned value was: None
[2023-04-04, 21:21:22 CST] {taskinstance.py:1288} INFO - Marking task as SUCCESS.
[2023-04-04, 21:21:22 CST] {local_task_job.py:154} INFO - Task exited with return
[2023-04-04, 21:21:22 CST] {local_task_job.py:264} INFO - 0 downstream tasks sche
```

後續讀者可以將 crawler，改成爬蟲程式，並設定 schedule_interval 後，就會定時爬蟲，並且可以查看歷史狀態、log，非常方便管理。

12.7.3 DummyOperator

用於建立一個，不做事情的 operator，主要是建立節點，如一個情境。例如某個股票策略，需要等股價、三大法人、外資持股三種資料，都完成爬蟲後，才能進行下一步策略計算，則使用 DummyOperator 節點，會更好管理。

我們來直接觀看範例，會比較好懂，如下。

crawler 爬蟲相關的任務，綁在一起，都執行完成後，再計算策略。

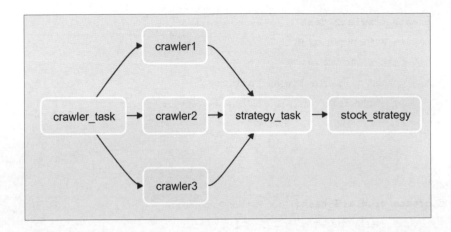

程式碼如下：

❏ etl/dummy_operator.py

https://github.com/FinMind/FinMindBook/blob/master/DataEngineering/
Chapter12/12.7/dataflow/etl/dummy_operator.py

```python
from airflow.operators.python_operator import (
    PythonOperator,
)


def create_crawler1_task() -> PythonOperator:
    return PythonOperator(
        task_id="crawler1",
        python_callable=lambda: print(
            "crawler1"
        ),
    )
```

```python
def create_crawler2_task() -> PythonOperator:
    return PythonOperator(
        task_id="crawler2",
        python_callable=lambda: print(
            "crawler2"
        ),
    )


def create_crawler3_task() -> PythonOperator:
    return PythonOperator(
        task_id="crawler3",
        python_callable=lambda: print(
            "crawler3"
        ),
    )


def create_stock_strategy_task() -> PythonOperator:
    return PythonOperator(
        task_id="stock_strategy",
        python_callable=lambda: print(
            "stock_strategy"
        ),
    )
```

❑ dags/dummy_operator.py

https://github.com/FinMind/FinMindBook/blob/master/DataEngineering/
Chapter12/12.7/dataflow/dags/dummy_operator.py

```
import airflow
from airflow.operators.dummy_operator import (
    DummyOperator,
)
from dataflow.constant import (
    DEFAULT_ARGS,
    MAX_ACTIVE_RUNS,
)
from dataflow.etl.dummy_operator import (
    create_crawler1_task,
    create_crawler2_task,
    create_crawler3_task,
    create_stock_strategy_task,
)

with airflow.DAG(
    dag_id="DummyOperator",
    default_args=DEFAULT_ARGS,
    schedule_interval=None,
    max_active_runs=MAX_ACTIVE_RUNS,
    catchup=False,
) as dag:
    crawler_task = DummyOperator(
        task_id="crawler_task"
    )
    strategy_task = DummyOperator(
        task_id="strategy_task"
    )
    (
        crawler_task
        >> [
            create_crawler1_task(),
```

```
        create_crawler2_task(),
        create_crawler3_task(),
    ]
    >> strategy_task
    >> create_stock_strategy_task()
)
```

主要架構是，先在 ETL 層，建立爬蟲、策略任務後，再到 DAG 層組裝相依性。

基本上，在小型架構上，使用到 DummyOperator 的機會不多，在大型資料流、需要有系統性管理的情況下，會有更多情境需使用 DummyOperator。

接著筆者介紹另一個 Operator，BranchPythonOperator。

12.7.4 BranchPythonOperator

根據此任務的回傳值，決定下一個步驟，要執行哪個任務。這樣描述有點抽象，我們直接來看下圖。

舉一個情境，當你在做爬蟲時，無法確定目標網站，什麼時候更新資料，因此一天當中，需要多次啟動爬蟲 DAG，並做事前檢查，Check Crawler，確認是否要進行爬蟲，則可以使用 BranchPython Operator。

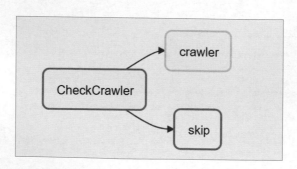

第一個任務，CheckCrawler，根據回傳值，判斷要執行 crawler 或是 skip 的任務。

程式碼如下：

❏ branch_python_operator.py

https://github.com/FinMind/FinMindBook/blob/master/DataEngineering/
Chapter12/12.7/dataflow/etl/branch_python_operator.py

```python
from airflow.operators.dummy_operator import (
    DummyOperator,
)
from airflow.operators.python_operator import (
    BranchPythonOperator,
    PythonOperator,
)


def create_skip_task() -> DummyOperator:
    return DummyOperator(task_id="skip")


def create_crawler_task() -> PythonOperator:
    return PythonOperator(
        task_id="crawler",
        python_callable=lambda: print(
            "crawler"
        ),
    )
```

```
def check_crawler():
    # 檢查今天是否以爬蟲
    # 如果已經爬蟲了，則跳過
    if True:
        return "skip"
    else:
        return "crawler"


def create_branch_python_operator_task():
    check_crawler_task = BranchPythonOperator(
        task_id="CheckCrawler",
        python_callable=check_crawler,
    )
    skip_task = create_skip_task()
    crawler_task = create_crawler_task()
    return check_crawler_task >> [
        skip_task,
        crawler_task,
    ]
```

這裡有一點非常特別，注意 check_crawler 函數，在此函數中，根據不同條件，決定回傳值，回傳的值非常重要，這跟 task_id 任務 ID 綁定。例如回傳 skip，就是告訴 Airflow，下一步執行 skip 任務，如果回傳 crawler，則下一步執行 crawler 任務。

本章節全部都使用 print 做範例，讀者可根據自己的使用場景，替換成不同的程式碼。

下一個場景，大多數開發 Airflow 的工程師，很常遇到的問題是，Airflow 環境與執行任務的 Python 環境有衝突，或是在 Airflow 管理多專案時，因為不同專案使用的 Python 環境不同，導致開多個 Airflow 去做排程，這讓管理分散了，理論上應該要統一環境，下一小節，筆者將介紹 DockerOperator，就是用於解決以上問題。

12.7.5 DockerOperator

顧名思義，建立一個，使用 Docker 容器為基礎，並執行相關指令的 Operator 任務。

（由於使用 DockerOperator，需額外套件，本節的 Pipfile 如下：

https://github.com/FinMind/FinMindBook/blob/master/DataEngineering/Chapter12/12.7/Pipfile

```
[[source]]
url = "https://pypi.org/simple"
verify_ssl = true
name = "pypi"

[packages]
apache-airflow = "==2.2.5"
pymysql = "==1.0.2"
dataflow = {editable = true, path = "."}
apache-airflow-providers-docker = "==3.0.0"

[dev-packages]

[requires]
python_version = "3.7"
```

增加了 apache-airflow-providers-docker）。

那要如何建立 DockerOperator 呢？以下是本書範例。

❏ dags/docker_operator.py

https://github.com/FinMind/FinMindBook/blob/master/DataEngineering/
Chapter12/12.7/dataflow/etl/docker_operator.py

```python
from airflow.operators.docker_operator import (
    DockerOperator,
)


def create_docker_operator_task() -> DockerOperator:
    return DockerOperator(
        task_id="DockerOperator",
        image="linsamtw/dataflow:12.7",
        command="pipenv run python dataflow/crawler.py",
        # 每次執行時,先拉取最新的 docker image
        force_pull=True,
        # 在退出容器時自動刪除容器。
        auto_remove=True,
    )
```

運行 crawler.py 函數,程式碼如下:

https://github.com/FinMind/FinMindBook/blob/master/DataEngineering/
Chapter12/12.7/dataflow/crawler.py

```python
def crawler_taiwan_stock_price():
    print("crawler_taiwan_stock_price")


if __name__ == "__main__":
    crawler_taiwan_stock_price()
```

DockerOperator 與其他的 Operator 不同，在此一一介紹參數：

- image="linsamtw/dataflow:12.7"
  - 顧名思義，使用 linsamtw/dataflow:12.7 這個 image 建立容器。
- command="pipenv run python dataflow/crawler.py"
  - 容器建立完成後，使用 command 的指令，執行程式。
- force_pull=True
  - 在每次執行任務時，強制拉取最新的 docker image。
  - 為什麼要這樣做呢？由於 docker image 可能會被更新，即使同一個版本，這是為了確保每次執行任務時，都是用最新的版本。
- auto_remove=True
  - 在執行完任務後，自動刪除容器。我們不希望遺留太多 container。

重點來了，這裡的 image 參數，可替換成任意的 docker image。也就是說，在同一個 Airflow 底下，即使執行不同的專案，也可以使用不同的 docker image，就不會造成 Python 版本、Package 衝突影響。完全由一個 Airflow，統一管理排程，後續不管要執行什麼任務，都被容器隔離，因此不受影響。

這裡有一點需特別注意，由於本書使用 Docker 假設 Airflow，如果要使用 DockerOperator，等於是在 Docker 環境下，去使用 Docker，相當複雜。這樣講可能有點抽象，可以直接看下圖。

最外層是 Docker，第一層是 Airflow Container 容器（因為使用 Docker 假設 Airflow），最後再到 Airflow 內，使用 DockerOperator。

容器包容器，變得非常複雜，所以筆者在此，對 airflow.yml 做了一些調整，如下。

https://github.com/FinMind/FinMindBook/blob/master/DataEngineering/Chapter12/12.7/airflow-docker-operator.yml

```
version: '3.0'
services:
  # ... 略過
  scheduler:
    image: linsamtw/dataflow:12.7
    hostname: "airflow-scheduler"
    command: pipenv run airflow scheduler
    depends_on:
      - initdb
    restart: always
    # swarm 設定
    deploy:
      mode: replicated
      replicas: 1
      placement:
        constraints: [node.labels.airflow == true]
    environment:
      - TZ=Asia/Taipei
    # 將容器內的 docker 與容器外的 docker 做連結
    volumes:
      - /var/run/docker.sock:/var/run/docker.sock
    networks:
      - my_network
```

```
networks:
  my_network:
    external: true
```

注意 scheduler 部分，增加了：

```
volumes:
  - /var/run/docker.sock:/var/run/docker.sock
```

這個意思是，將容器內的 docker 與容器外的 docker 做連結，讓 Airflow 在執行 DockerOperator 時，使用的 Docker 指令，會跟容器外做溝通、連結，這樣講可能還是太抽象，直接看下圖。

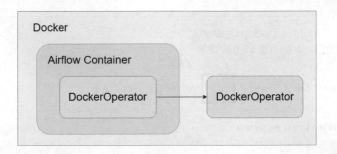

由於 Airflow 容器內外相通，這讓 Airflow 在執行 DockerOperator 時，會到容器外，建立一個 Docker 容器，並執行相關指令，讀者這時可以跳轉到 http://localhost:8888/，執行 DockerOperator，如下：

按照 1、2 步驟，執行 DockerOperator 這個 DAG。

執行完後，會出現圖 3。

如果任務成功執行後，查看 log，如下：

```
[2023-04-05, 02:47:05 CST] {taskinstance.py:1270} INFO - Executing <Task(DockerOperator): DockerOperator> on 2023-04-0
[2023-04-05, 02:47:05 CST] {standard_task_runner.py:52} INFO - Started process 400 to run task
[2023-04-05, 02:47:05 CST] {standard_task_runner.py:79} INFO - Running: ['airflow', 'tasks', 'run', 'DockerOperator',
[2023-04-05, 02:47:05 CST] {standard_task_runner.py:80} INFO - Job 30: Subtask DockerOperator
[2023-04-05, 02:47:05 CST] {logging_mixin.py:109} INFO - Running <TaskInstance: DockerOperator.DockerOperator manual_
[2023-04-05, 02:47:05 CST] {taskinstance.py:1448} INFO - Exporting the following env vars:
AIRFLOW_CTX_DAG_OWNER=FinMind
AIRFLOW_CTX_DAG_ID=DockerOperator
AIRFLOW_CTX_TASK_ID=DockerOperator
AIRFLOW_CTX_EXECUTION_DATE=2023-04-04T18:47:03.002375+00:00
AIRFLOW_CTX_DAG_RUN_ID=manual__2023-04-04T18:47:03.002375+00:00
[2023-04-05, 02:47:05 CST] {docker.py:373} INFO - Pulling docker image linsamtw/dataflow:12.7
[2023-04-05, 02:47:09 CST] {docker.py:387} INFO - 12.7: Pulling from linsamtw/dataflow
[2023-04-05, 02:47:09 CST] {docker.py:382} INFO - Digest: sha256:8fe3ac9499be570dd42f3bfd53ebd10ce113a3452dc9e745243f(
[2023-04-05, 02:47:09 CST] {docker.py:382} INFO - Status: Image is up to date for linsamtw/dataflow:12.7
[2023-04-05, 02:47:09 CST] {docker.py:248} INFO - Starting docker container from image linsamtw/dataflow:12.7
[2023-04-05, 02:47:09 CST] {docker.py:259} WARNING - Using remote engine or docker-in-docker and mounting temporary v
[2023-04-05, 02:47:10 CST] {docker.py:310} INFO - crawler_taiwan_stock_price
[2023-04-05, 02:47:10 CST] {taskinstance.py:1288} INFO - Marking task as SUCCESS. dag_id=DockerOperator, task_id=Dock
[2023-04-05, 02:47:10 CST] {local_task_job.py:154} INFO - Task exited with return code 0
[2023-04-05, 02:47:10 CST] {local_task_job.py:264} INFO - 0 downstream tasks scheduled from follow-on schedule check
```

可以看到：

圖 5，拉取 linsamtw/dataflow:12.7。

圖 6，啟動 container。

圖 7，執行該程式（此程式只印出 crawler_taiwan_stock_price，後續讀者可自行更改成爬蟲程式）。

以上，就完成建立 BashOperator、PythonOperator、DummyOperator、BranchPythonOperator、DockerOperator 了，基本上可以應付 80% 的情境了。

本章節到此，都只介紹，單機版本的 Airflow，下一章節，將結合 Celery，並與爬蟲做結合。

12.8 Airflow 結合爬蟲 - CeleryExecutor

本小節,將使用 Airflow 結合 CeleryExecutor,並借用 Chapter5 的爬蟲、上傳資料庫,實現用 Airflow 管理排程的爬蟲架構。

　　　　那跟過去提到的,**APScheduler** 相比,有什麼好處呢?

Airflow 有完善的介面、能讓你監控多個爬蟲狀態、留存 log,甚至在爬蟲失敗時,有辦法重啟。不論是筆者在業界的經驗,或是 FinMind 的架構,都有超過 50 個、100 個排程要管理,有介面統一管理,非常重要,本書將會一步步揭露整個架構,讓讀者在本書中,就能接觸業界的真實架構。

12.8.1 架構介紹

首先介紹本節架構,如下:

https://github.com/FinMind/FinMindBook/tree/master/DataEngineering/Chapter12/12.8

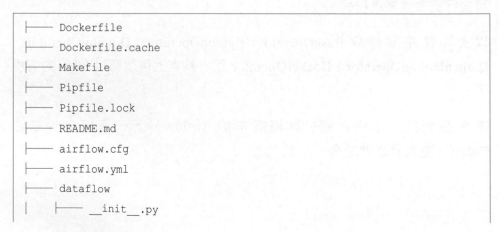

```
├── Dockerfile
├── Dockerfile.cache
├── Makefile
├── Pipfile
├── Pipfile.lock
├── README.md
├── airflow.cfg
├── airflow.yml
├── dataflow
│   ├── __init__.py
```

```
|   ├── backend
|   |   ├── __init__.py
|   |   └── db
|   |       ├── __init__.py
|   |       ├── clients.py
|   |       ├── db.py
|   |       └── router.py
|   ├── config.py
|   ├── constant.py
|   ├── crawler
|   |   ├── __init__.py
|   |   └── taiwan_stock_price.py
|   ├── dags
|   |   ├── __init__.py
|   |   └── taiwan_stock_price.py
|   ├── etl
|   |   ├── __init__.py
|   |   └── taiwan_stock_price.py
|   └── schema
|       ├── __init__.py
|       └── dataset.py
├── genenv.py
├── local.ini
├── mysql.yml
├── portainer.yml
└── setup.py
```

除了常見的 Pipfile、Dockerfile、Makefile 等設定檔以外，dataflow 底下
的 backend、crawler、schema 資料夾，就是筆者從 5.5.5 複製過來的，
並做一些小調整，整體不變。而 dataflow 底下的 dags、etl 資料夾，就是
12.7 介紹過的。

由於節合 Airflow 與 Celery，Pipfile 增加幾個 Package，如下：

https://github.com/FinMind/FinMindBook/blob/master/DataEngineering/
Chapter12/12.8/Pipfile

```
[[source]]
url = "https://pypi.org/simple"
verify_ssl = true
name = "pypi"

[packages]
apache-airflow = "==2.2.5"
pymysql = "==1.0.2"
dataflow = {editable = true, path = "."}
apache-airflow-providers-docker = "==3.0.0"
apache-airflow-providers-celery = "==3.0.0"
flower = "==1.0.0"
importlib-metadata = "==4.13.0"
celery = {extras = ["redis"], version = "*"}
pydantic = "==1.8.2"
pandas = "==1.1.5"
loguru = "==0.5.3"


[dev-packages]


[requires]
python_version = "3.7"
```

由於使用 CeleryExecutor，因此需要新增幾個 Package。celery、flower 就
不用多說了，這是 celery 所需的套件，其餘 pydantic、pandas、loguru 就
跟 Chapter 5 一樣。

另一個不同的地方是，airflow.cfg，因為要使用 CeleryExecutor，做了一些調整：

1. 在 core 部分，將 executor 設定成 CeleryExecutor。

https://github.com/FinMind/FinMindBook/blob/master/DataEngineering/
Chapter12/12.8/airflow.cfg#L1

```
[core]
dags_folder = /FinMindProject/dataflow/dags
hostname_callable = socket.getfqdn
default_timezone = Asia/Taipei
executor = CeleryExecutor
```

2. 在 celery 部分，設定 broker_url、result_backend、flower_host：

    a. broker_url：本節選用 Redis，而非 RabbitMQ，原因是需要 result_backend。

    b. result_backend：如 12.5.3 所說，需要將任務的結果，存到某個地方。本節選用 Redis，因為效能夠快，一般資料庫都是存在硬碟，但 Redis 是存到記憶體。這也是為什麼 borker_url 也選用 Redis，效能考量。

    c. flower_host：設定 host。在 5.5.2 建立分散式架構時，使用 Flower 監控工人狀態，因此在此，也需要 Flower。

https://github.com/FinMind/FinMindBook/blob/master/DataEngineering/
Chapter12/12.8/airflow.cfg#L196

```
[celery]
celery_app_name = airflow.executors.celery_executor
worker_concurrency = 16
worker_umask = 0o077
```

```
broker_url = redis://redis:6379/0
result_backend = redis://redis:6379/1
flower_host = flower
flower_url_prefix =
flower_port = 5555
```

調整完 airflow.cfg 後，下一步是 airflow.yml。

https://github.com/FinMind/FinMindBook/blob/master/DataEngineering/
Chapter12/12.8/airflow.yml

```
version: '3.8'
services:
  initdb:
    image: linsamtw/dataflow:12.8
    command: pipenv run airflow db init
    restart: on-failure
    # swarm 設定
    deploy:
      mode: replicated
      replicas: 1
      placement:
        constraints: [node.labels.airflow == true]
    networks:
        - my_network

  create-user:
    image: linsamtw/dataflow:12.8
    command: pipenv run airflow users create --username admin --firstname
lin --lastname sam --role Admin -p admin --email finmind.tw@gmail.com
    depends_on:
      - initdb
    restart: on-failure
```

```
  # swarm 設定
  deploy:
    mode: replicated
    replicas: 1
    placement:
      constraints: [node.labels.airflow == true]
  networks:
      - my_network

redis:
  image: 'bitnami/redis:5.0'
  ports:
      - 6379:6379
  volumes:
      - 'redis_data:/bitnami/redis/data'
  environment:
      - ALLOW_EMPTY_PASSWORD=yes
  restart: always
  # swarm 設定
  deploy:
    mode: replicated
    replicas: 1
    placement:
      constraints: [node.labels.airflow == true]
  networks:
      - my_network

webserver:
  image: linsamtw/dataflow:12.8
  hostname: "airflow-webserver"
  command: pipenv run airflow webserver -p 8888
  depends_on:
```

```
    - initdb
  restart: always
  ports:
    - 8888:8888
  environment:
   - TZ=Asia/Taipei
  # swarm 設定
  deploy:
   mode: replicated
   replicas: 1
   placement:
     constraints: [node.labels.airflow == true]
  networks:
    - my_network

flower:
  image: mher/flower:0.9.5
  restart: always
  depends_on:
    - redis
  command: ["flower", "--broker=redis://redis:6379/0", "--port=5555"]
  ports:
    - "5555:5555"
  # swarm 設定
  deploy:
   mode: replicated
   replicas: 1
   placement:
     constraints: [node.labels.airflow == true]
  networks:
    - my_network
```

```
scheduler:
  image: linsamtw/dataflow:12.8
  hostname: "airflow-scheduler"
  command: pipenv run airflow scheduler
  depends_on:
    - initdb
  restart: always
  environment:
    - TZ=Asia/Taipei
  volumes:
    - /var/run/docker.sock:/var/run/docker.sock
  # swarm 設定
  deploy:
    mode: replicated
    replicas: 1
    placement:
      constraints: [node.labels.airflow == true]
  networks:
      - my_network

worker:
  image: linsamtw/dataflow:12.8
  hostname: "{{.Service.Name}}.{{.Task.Slot}}"
  restart: always
  depends_on:
      - scheduler
  command: pipenv run airflow celery worker
  # swarm 設定
  deploy:
    mode: replicated
    replicas: 3
    placement:
```

```
        constraints: [node.labels.worker == true]
    networks:
        - my_network

crawler_twse:
    image: linsamtw/dataflow:12.8
    hostname: "{{.Service.Name}}.{{.Task.Slot}}"
    restart: always
    depends_on:
        - scheduler
    command: pipenv run airflow celery worker -q twse
    # swarm 設定
    deploy:
      mode: replicated
      replicas: 1
      placement:
        max_replicas_per_node: 1
        constraints: [node.labels.crawler_twse == true]
    networks:
        - my_network

crawler_tpex:
    image: linsamtw/dataflow:12.8
    hostname: "{{.Service.Name}}.{{.Task.Slot}}"
    restart: always
    depends_on:
        - scheduler
    command: pipenv run airflow celery worker -q tpex
    # swarm 設定
    deploy:
      mode: replicated
      replicas: 1
```

```
    placement:
      max_replicas_per_node: 1
      constraints: [node.labels.crawler_tpex == true]
    networks:
      - my_network


networks:
  my_network:
    external: true


volumes:
  redis_data:
```

可以看到，做了非常多調整，基本上 Airflow 要改成 CeleryExecutor
分散式架構，非常繁瑣，一般在小型專案、甚至小公司，是不會走到
CeleryExecutor 的，本書秉持著分享業界經驗出發，希望分享進階的技
能。

以下將介紹這份 airflow.yml 新增了哪些服務：

1. redis

    a. 如前幾頁所說，需要 Redis 當作 broker_url、result_backend，因此
       在 airflow.yml 新增 Redis 服務。

```
redis:
  image: 'bitnami/redis:5.0'
  ports:
    - 6379:6379
  volumes:
    - 'redis_data:/bitnami/redis/data'
  environment:
```

```
     - ALLOW_EMPTY_PASSWORD=yes
restart: always
networks:
    - my_network
```

2. flower

   a. 本節 Airflow 也需要 Flower，特別注意一點，因為使用 Redis 當作 Borker。

   b. 這裡設定 broker=redis://redis:6379/0。

```
flower:
  image: mher/flower:0.9.5
  restart: always
  depends_on:
      - redis
  command: ["flower", "--broker=redis://redis:6379/0", "--port=5555"]
  ports:
      - "5555:5555"
  networks:
      - my_network
```

3. worker

   a. 如同 Chapter 5 中，工人的腳色。

   b. hostname：過去 hostname 都直接寫定成固定的值，但想像一下，如果有多個工人，那名字都一樣，當有一個工人出錯時，不容易去判斷到底是誰。如下圖，工人名字增加了編號 1、2、3，讓監控上更方便。

c. command：在 Airflow 架構下，啟動工人的指令。

d. 其餘參數設定，都與 Chapter 5 類似，就不再重複敘述了。

```
worker:
  image: linsamtw/dataflow:12.8
  hostname: "{{.Service.Name}}.{{.Task.Slot}}"
  restart: always
  depends_on:
      - scheduler
  command: pipenv run airflow celery worker
  # swarm 設定
  deploy:
    mode: replicated
    replicas: 3
    placement:
      constraints: [node.labels.worker == true]
  networks:
      - my_network
```

4. crawler_twse

    a. 如同 Chapter 5 中，工人的腳色，且只處理 twse 相關的任務。

    b. command：該服務只處理 twse 的任務，設定上跟 celery 些微不同。

    c. max_replicas_per_node：顧名思義，讓每個 node 機器，最多只能有一個 crawler_twse 的服務。目的是不希望同一台機器，啟動太多 crawler_twse，這會導致同一個 IP，同一時間大量爬蟲，容易被 ban。

```
crawler_twse:
  image: linsamtw/dataflow:12.8
  hostname: "{{.Service.Name}}.{{.Task.Slot}}"
  restart: always
  depends_on:
      - scheduler
  command: pipenv run airflow celery worker -q twse
  # swarm 設定
  deploy:
    mode: replicated
    replicas: 1
    placement:
      max_replicas_per_node: 1
      constraints: [node.labels.crawler_twse == true]
  networks:
      - my_network
```

5. crawler_tpex

    a. 與 crawler_twse 類似，就不再多做贅述。

```
crawler_tpex:
  image: linsamtw/dataflow:12.8
  hostname: "{{.Service.Name}}.{{.Task.Slot}}"
  restart: always
```

```
depends_on:
    - scheduler
command: pipenv run airflow celery worker -q tpex
# swarm 設定
deploy:
  mode: replicated
  replicas: 1
  placement:
    max_replicas_per_node: 1
    constraints: [node.labels.crawler_tpex == true]
networks:
    - my_network
```

看完以上 yml 後，有發現什麼嗎？在 Airflow + CeleryExecutor 架構下，可自由調整工人的數量 replicas，讓整個架構，大大增加了拓展性，就如同 Chapter 5 介紹的概念，讓未來讀者開發的 Airflow，能輕易地擴充，這也是為什麼當初要使用 Docker 架設的原因之一，這才是業界在使用 Airflow 的方式。

本書在此，介紹了如何將 Airflow 從單機版，擴充到分散式版本，下一小節，將進行實際架設。

12.8.2 架設 Airflow + CeleryExecutor

1. 步驟 1，先清空原本的 Airflow。

```
docker stack rm airflow
```

2. 步驟 2，部屬 Airflow。

```
docker stack deploy -c airflow.yml airflow
```

部屬完成後，可以到 Portainer 頁面 http://localhost:9000/ 查看，如下：

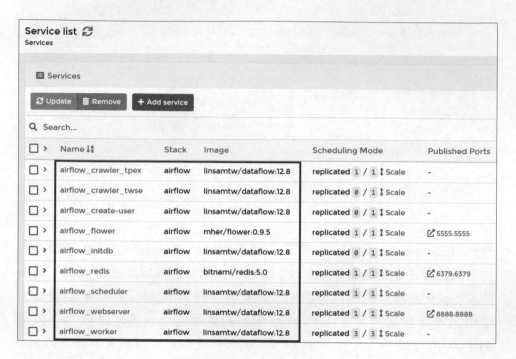

"""" 小提醒，如果出現下圖，正在 **pending** 狀態 """"

請讀者到 Swarm 的地方：

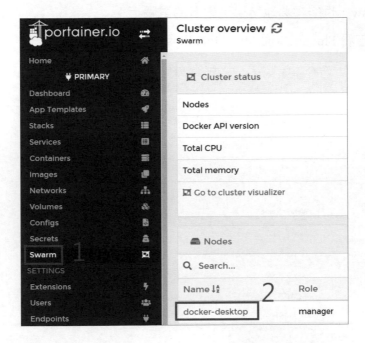

做以下設定：

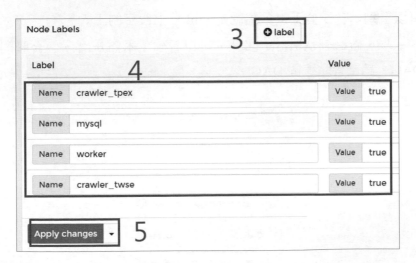

3. 以上就部屬完成啦！是不是非常簡單，因為本書使用 Docker 架設 Airflow，也能輕易地移植到讀者的電腦上。

4. 接著查看 Flower 頁面 http://localhost:5555/。

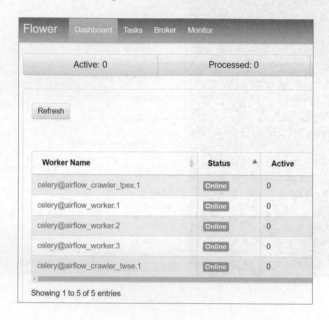

可以看到，有 5 個工人，這就是 yml 檔上設定的工人。

5. 下一步，跳轉到 Airflow 部分，可以看到，本節新增一個台股爬蟲的 DAG，請讀者點這個 DAG。

6. 接著點選 Graph，這個圖比較好理解。

7. 之後如下圖，可以看到，台股股價的兩個爬蟲，twse - 證交所、tpex - 櫃買中心。

8. 程式碼部分，會留到下一個小節講解，在此，先使用 config 驅動 DAG（Trigger DAG w/ config ）進行爬蟲。

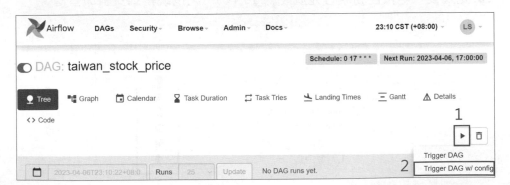

9. 輸入你要爬蟲的日期，在此範例輸入 2023-03-28，之後按下 Trigger
   啟動。

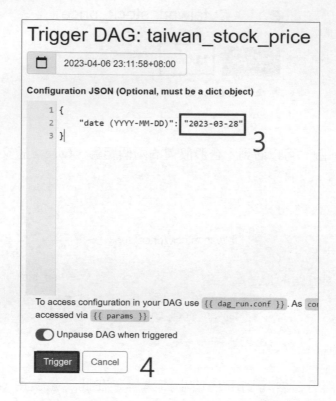

10. 完成後，等待任務執行完成（約 10 秒），之後跳轉到資料庫查看。

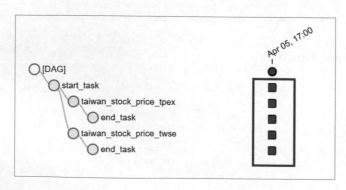

11. 跳轉到資料庫 http://localhost:8080/，如下圖，（還記得在 Chapter 5.4 節，建立 financialdata 資料庫，並建立 TaiwanStockPrice 表嗎？如果 還沒做的讀者，麻煩先建立表，才能成功上傳資料喔）可以看到， TaiwanStockPrice 表中，已經有資料囉，代表我們使用 Airflow，成功 爬蟲，並上傳資料到資料庫。

接著，讀者是不是很好奇，實際上程式碼怎麼撰寫，下一小節筆者將解 析程式碼。

12.8.3 程式碼解析

撇除一些介紹過的 yml、cfg、config 等設定檔，直接看來 Python 架構。

https://github.com/FinMind/FinMindBook/tree/master/DataEngineering/ Chapter12/12.8

```
├── dataflow
│   ├── __init__.py
│   ├── backend
│   │   ├── __init__.py
│   │   └── db
│   │       ├── __init__.py
│   │       ├── clients.py
│   │       ├── db.py
│   │       └── router.py
│   ├── config.py
│   ├── constant.py
│   ├── crawler
│   │   ├── __init__.py
│   │   └── taiwan_stock_price.py
│   ├── dags
│   │   ├── __init__.py
│   │   └── taiwan_stock_price.py
│   ├── etl
│   │   ├── __init__.py
│   │   └── taiwan_stock_price.py
│   └── schema
│       ├── __init__.py
│       └── dataset.py
```

基本上 backend、schema、crawler 都是直接複製 5.5.5，不多做介紹，直接進入核心程式。

首先是進入點，DAG：

❏ dags/taiwan_stock_price.py

https://github.com/FinMind/FinMindBook/blob/master/DataEngineering/
Chapter12/12.8/dataflow/dags/taiwan_stock_price.py

```python
import airflow

from dataflow.constant import (
    DEFAULT_ARGS,
    MAX_ACTIVE_RUNS,
)
from dataflow.etl.taiwan_stock_price import (
    create_crawler_taiwan_stock_price_task,
)
from airflow.operators.dummy_operator import (
    DummyOperator,
)

with airflow.DAG(
    dag_id="taiwan_stock_price",
    default_args=DEFAULT_ARGS,
    # 設定每天 17:00 執行爬蟲
    schedule_interval="0 17 * * *",
    max_active_runs=MAX_ACTIVE_RUNS,
    # 設定參數，Airflow 除了按按鈕，單純的執行外
    # 也可以在按按鈕時，帶入特定參數
    # 在此設定 date 參數，讓讀者可自行輸入，想要爬蟲的日期
    params={
        "date (YYYY-MM-DD)": "",
    },
    catchup=False,
) as dag:
    start_task = DummyOperator(
        task_id="start_task"
    )
    end_task = DummyOperator(
        task_id="end_task"
```

```
)
crawler_taiwan_stock_price_task = (
    create_crawler_taiwan_stock_price_task()
)
(
    start_task
    >> crawler_taiwan_stock_price_task
    >> end_task
)
```

如註解所說：

- schedule_interval：設定每天定時爬蟲時間點，股價資料大約在 1700 會更新完成，因此訂在這時段，讀者可根據自己的情境設定。
- params：設定這個 DAG，可以用參數輸入的方式啟動。

而以下這段 data-pipeline 任務流程，可以直接對照 DAG 圖。

```
(
    start_task
    >> crawler_taiwan_stock_price_task
    >> end_task
)
```

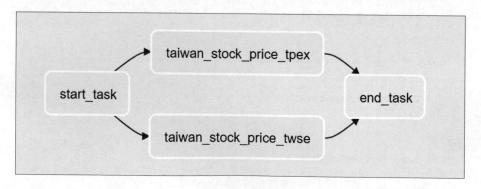

基本上 DAG 很單純,負責設定介面的參數、多久執行一次,資料流怎麼走等,而任務如何執行,則是 ETL 層負責,如下:

❏ etl/taiwan_stock_price.py

https://github.com/FinMind/FinMindBook/blob/master/DataEngineering/Chapter12/12.8/dataflow/etl/taiwan_stock_price.py

```
import datetime

from airflow.operators.python_operator import (
    PythonOperator,
)

from dataflow.backend import db
from dataflow.crawler.taiwan_stock_price import (
    crawler,
)


def crawler_taiwan_stock_price_twse(
    **kwargs,
):
    # 由於在 DAG 層,設定 params,可輸入參數
    # 因此在此,使用以下 kwargs 方式,拿取參數
    # DAG 中 params 參數設定是 date (YYYY-MM-DD)
    # 所以拿取時,也要用一樣的字串
    params = kwargs["dag_run"].conf
    date = params.get(
        "date (YYYY-MM-DD)",
        # 如果沒有帶參數,則預設 date 是今天
        datetime.datetime.today().strftime(
```

```
            "%Y-%m-%d"
        ),
    )
    # 進行爬蟲
    df = crawler(
        dict(
            date=date,
            data_source="twse",
        )
    )
    # 資料上傳資料庫
    db.upload_data(
        df,
        "TaiwanStockPrice",
        db.router.mysql_financialdata_conn,
    )


def crawler_taiwan_stock_price_tpex(
    **kwargs,
):
    # 註解如上
    params = kwargs["dag_run"].conf
    date = params.get(
        "date (YYYY-MM-DD)",
        datetime.datetime.today().strftime(
            "%Y-%m-%d"
        ),
    )
    df = crawler(
        dict(
            date=date,
```

```
                data_source="tpex",
        )
    )
    db.upload_data(
        df,
        "TaiwanStockPrice",
        db.router.mysql_financialdata_conn,
    )



def create_crawler_taiwan_stock_price_task() -> PythonOperator:
    return [
        # 建立任務
        PythonOperator(
            task_id="taiwan_stock_price_twse",
            python_callable=crawler_taiwan_stock_price_twse,
            queue="twse",
            provide_context=True,
        ),
        PythonOperator(
            task_id="taiwan_stock_price_tpex",
            python_callable=crawler_taiwan_stock_price_tpex,
            queue="tpex",
            provide_context=True,
        ),
    ]
```

大多數改動，都有註解可以參考：

1. 在 DAG 設定 params，那要如何拿取呢？方法如下：

```
params = kwargs["dag_run"].conf
date = params.get(
```

```
    "date (YYYY-MM-DD)",
    # 如果沒有帶參數，則預設 date 是今天
    datetime.datetime.today().strftime(
        "%Y-%m-%d"
    ),
)
```

2. 接著進行爬蟲

```
df = crawler(
    dict(
        date=date,
        data_source="twse",
    )
)
```

3. 最後上傳資料庫

```
db.upload_data(
    df,
    "TaiwanStockPrice",
    db.router.mysql_financialdata_conn,
)
```

由於在 5.5.5，已經實作了爬蟲程式碼，因此在 ETL 層，只是單純建立任務，帶入參數即可。基本上只有改動到 DAG、ETL 層，就完成了，是不是非常簡單呢？讀者後續可使用此架構，進一步做延伸。

本節的架構，達到良好的職責分離。

- DAG 負責建立資料流、介面、管理任務。
- ETL 負責建立任務，並帶入參數。
- Crawler 層，負責實作爬蟲。

12.9 結論

最後來進行總結，首先介紹了什麼是排程管理工具 - Airflow，為什麼選用 Airflow，並使用 Docker 成功架設第一個 Airflow，也解釋了為什麼選用 Docker。接著介紹了 DAG、各種 Operator，最後結合 Celery、爬蟲，使用 Airflow 進行分散式爬蟲。

想像一下，在 Chapter 7 使用 linode 申請多台機器，分散式提高效能，而 Airflow 結合 Celery 後，也能達到分散式效果，未來效能將不會是瓶頸，這也是業界與學習上的差別，高效能的架構、系統。

到此，完整展現，Airflow 合併 Celery、爬蟲，等於是融合本書各個章節。到此可以告訴自己，自己真的會 Airflow 了。單純跑跑 Airflow 範例，並不代表學會，能將新學習的技術，與過去融合，並在真實場景做應用，才是真的「會」。在業界也是，大多數情境，是在現有的架構上，導入新技術，這 時如果單純會 Airflow，但沒有能力與其他架構結合，那也是無法應用的。

用 Docker 來建置 Redis 的 Delev

13.1　什麼是 Redis？

Redis 是一種基於記憶體的資料庫,那跟一般資料庫有什麼差別呢?多數
資料庫都是將資料寫入硬碟,而 Redis 是寫入記憶體,硬碟與記憶體相
比,兩者效能天差地遠,這就是 Redis 最大的特色。

那 Redis 在業界,大多使用在哪些場景呢?對於需要做快取的工作,大多
數都會選用 Redis 或是與 Redis 類似,基於記憶體的資料庫。例如一般在
訪問網站時,網站回傳的數據,都是基於 Redis。

想像一下,如果直接對資料庫做寫入、讀取,會發生什麼事情?當網站
一秒有 1,000 個用戶登入時,那系統就會對資料庫做 1,000 次讀取,這對

資料庫來說，負擔非常大，但是如果先把資料存到 Redis，接著對 Redis 做 1,000 次讀取，將大大減輕資料庫負擔。而 Redis 是基於記憶體的資料庫，記憶體讀寫效能，比硬碟快 100 倍、1,000 倍，這就是為什麼需要使用 Redis 做快取。

另一個場景，在做大數據，即時分析系統、即時預測系統，對於效能有非常高的需求，這時就會使用 Redis 來進行資料的存取。例如用戶登入 YouTube，YouTube 會基於推薦系統，推薦你不同的影片，這時的推薦結果，基本上都是模型做完預測後，存進 Redis 的。

介紹了這麼多，不如實際操作，更能了解 Redis，下一個小節，將使用本書最常用的工具 - Docker 進行架設。

13.2　使用 Docker 架設 Redis - 結合 Celery

本節架構如下，基本上是複製 5.5.3，只是把 rabbitmq.yml 改成 redis.yml，筆者在此，將展示 Redis 結合 Celery，作為範例。

https://github.com/FinMind/FinMindBook/tree/master/DataEngineering/Chapter13

```
├── Makefile
├── Pipfile
├── Pipfile.lock
├── README.md
├── producer.py
├── redis.yml
├── tasks.py
└── worker.py
```

當你的分散式系統，每天有上萬個任務需要被執行，遇到 RabbitMQ 效能不足時，或許可以考慮 Redis，Redis 效能更強。

那要如何架設呢？首先，yml 檔如下，將使用這份 yml 檔進行架設。

https://github.com/FinMind/FinMindBook/blob/master/DataEngineering/Chapter13/redis.yml

```yml
version: '3'
services:

  redis:
    image: 'bitnami/redis:5.0'
    ports:
        - 6379:6379
    volumes:
        - 'redis_data:/bitnami/redis/data'
    environment:
        - ALLOW_EMPTY_PASSWORD=yes
    restart: always
    networks:
        - dev


  flower:
    image: mher/flower:0.9.5
    command: ["flower", "--broker=redis://redis:6379/0", "--port=5555"]
    ports: # docker publish port 5555 to 5555 (將 docker 內部 ip 5555,
跟外部 5555 做連結)
      - 5555:5555
    depends_on:
      - redis
    networks:
      - dev
```

```
networks:
  dev:


volumes:
  redis_data:
```

Redis 的 Port 與 RabbitMQ 不同，在 6379，因此 ports 的設定上，改成 6379。同樣 Flower 原先是對 RabbitMQ 做連線，現在改成對 Redis 做連線，因此 Command 也做更動。

接著，使用以下指令進行架設（相信讀者在此，都很熟悉了）。

```
docker-compose -f redis.yml up -d
```

架設完成後，使用與 5.5.3 相同的指令，啟動工人 Worker。

```
pipenv run celery -A worker worker --loglevel=info
```

結果如下圖。

```
sam@DESKTOP-IKT69L5:~/FinMindBook/DataEngineering/Chapter13$ pipenv run celery -A worker worker --loglevel=info

 -------------- celery@DESKTOP-IKT69L5 v5.1.2 (sun-harmonics)
--- ***** -----
-- ******* ---- Linux-4.4.0-19041-Microsoft-x86_64-with-Ubuntu-18.04-bionic 2023-04-07 02:00:19
- *** --- * ---
- ** ---------- [config]
- ** ---------- .> app:         task:0x7f17c8903d30
- ** ---------- .> transport:   redis://localhost:6379/0
- ** ---------- .> results:     disabled://
- *** --- * --- .> concurrency: 12 (prefork)
-- ******* ---- .> task events: OFF (enable -E to monitor tasks in this worker)
--- ***** -----
 -------------- [queues]
                .> celery           exchange=celery(direct) key=celery


[tasks]
  . tasks.crawler

[2023-04-07 02:00:20,071: INFO/MainProcess] Connected to redis://localhost:6379/0
[2023-04-07 02:00:20,079: INFO/MainProcess] mingle: searching for neighbors
[2023-04-07 02:00:21,106: INFO/MainProcess] mingle: all alone
[2023-04-07 02:00:21,134: INFO/MainProcess] celery@DESKTOP-IKT69L5 ready.
[2023-04-07 02:00:24,015: INFO/MainProcess] Events of group {task} enabled by remote.
```

成功後，試著發送任務，指令與 5.5.3 相同。

```
pipenv run python producer.py
```

成功發送後，查看工人 Worker 狀態。

如果結果如下圖，那代表成功了。工人成功接到任務，並將爬蟲、上傳
資料庫的字串，印出來，與 5.5.3 相同。

```
[tasks]
  . tasks.crawler

[2023-04-07 02:00:20,071: INFO/MainProcess] Connected to redis://localhost:6379/0
[2023-04-07 02:00:20,079: INFO/MainProcess] mingle: searching for neighbors
[2023-04-07 02:00:21,106: INFO/MainProcess] mingle: all alone
[2023-04-07 02:00:21,134: INFO/MainProcess] celery@DESKTOP-IKT69L5 ready.
[2023-04-07 02:00:24,015: INFO/MainProcess] Events of group {task} enabled by remote.
[2023-04-07 02:02:30,458: INFO/MainProcess] Task tasks.crawler[81bd2d72-c4ec-4b6d-9948-e75e43aaf23f] received
[2023-04-07 02:02:30,461: WARNING/ForkPoolWorker-8] crawler
[2023-04-07 02:02:30,462: WARNING/ForkPoolWorker-8]

[2023-04-07 02:02:30,462: WARNING/ForkPoolWorker-8] upload db
[2023-04-07 02:02:30,462: WARNING/ForkPoolWorker-8]
```

到這就完成了，使用 Redis 替換 RabbitMQ，是不是很簡單呢？讀者這時
可能會想問，程式碼需要做調整嗎？有兩個地方需要做調整：

1. Pipfile，由於使用 Redis，因次 Celery 要多安裝一個 Redis 套件，安
   裝方法如下。

```
pipenv install celery[redis]
```

https://github.com/FinMind/FinMindBook/blob/master/DataEngineering/
Chapter13/Pipfile

```
[[source]]
url = "https://pypi.org/simple"
verify_ssl = true
name = "pypi"
```

```
[packages]
celery = {extras = ["redis"], version = "*"}

[dev-packages]

[requires]
python_version = "3.6"
```

2. worker.py，工人的 Broker 設定，從對 RabbitMQ 連線，改成對 Redis
   連線。

https://github.com/FinMind/FinMindBook/blob/master/DataEngineering/
Chapter13/worker.py

```
from celery import Celery


app = Celery(
    "task",
    # 只包含 tasks.py 裡面的程式，才會成功執行
    include=["tasks"],
    # 連線到 redis,
    broker="redis://localhost:6379/0",
)
```

就只需做這兩個調整即可，其餘程式碼、指令，都不需要更動，就能簡
單的提高效能。

實際上 Redis 有非常多功能，但本書著重在分散式部分，因此本節，單純
介紹 Redis 如何與 Celery 結合，避免偏離主題。

下一個章節，算是最多讀者來信詢問的 --- 監控系統。基本上在業界、
產品開發上，一定會做監控系統，監控各種服務的穩定度，以下將公開
FinMind 的監控系統架構。

第 6 篇
監控系統

監控工具介紹

14.1 為什麼需要監控系統？

顧名思義，幫助產品開發、維運上，做系統監控。

本書到此，架設了分散式爬蟲系統、MySQL 資料庫、API 接口、視覺化工具 Redash、排程管理工具 Airflow，這 5 個主要的系統，架設多個系統，勢必需要多台雲端機器。而監控系統，能大大幫助工程師做管理、維運系統、維運雲端機器，提高效率（基本上一個產品，一定會用到多種系統）。

可能會有讀者想問，MySQL & phpMyAdmin 本身有內建監控系統：

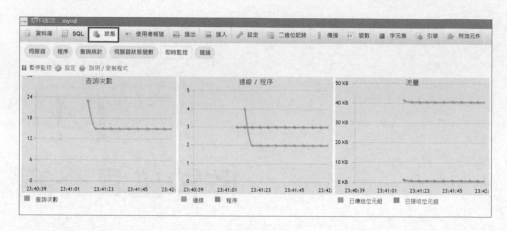

而 RabbitMQ 也有監控系統：

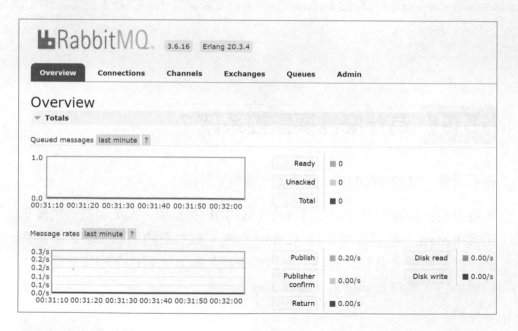

當然 Airflow 等系統也有內建的監控系統,那為什麼要使用監控系統呢?

因為監控系統,目的是 """ 統一監控 """,在同一個平台,監控多個系統。當一個產品架設許多系統後(如 FinMind),不太可能去 3、5 個不同地方,觀看系統狀態,有一個 """ 統一監控 """ 系統,能大大幫助維運,當然除了統一監控以外,還有其他優勢,本書接下來將一一介紹。

14.2 最知名的開源監控系統之一

一個產品,需要做到許多監控,本書將介紹最常見的開源監控系統之一,Grafana & Prometheus,這是台灣業界真實在使用的工具。

讀者有興趣,可以上網查詢相關職缺,技能需求上,大多都會寫,希望有 Grafana & Prometheus 經驗,如華碩、新光金、中華電信、聯發科、台積電、蝦皮、17LIVE 等等(以上資訊來自 google 相關職缺)。

筆者先展示基礎的 Grafana 監控系統。

1. CPU 用量監控

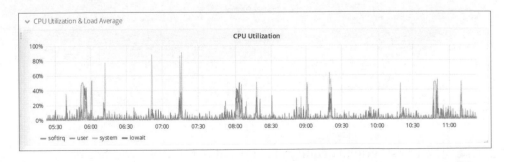

2. 記憶體用量監控

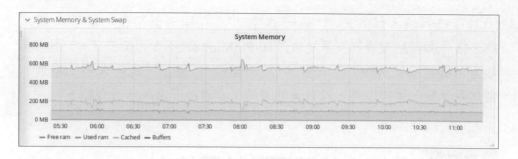

3. 硬碟用量監控

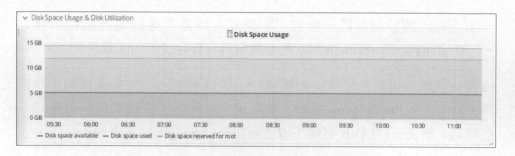

4. 網路流量監控

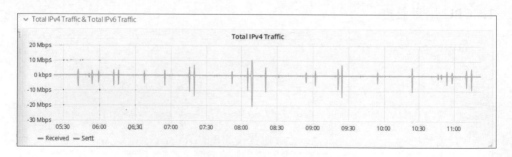

（上圖引用 https://grafana.com/grafana/dashboards/1295-yourhostname/）

以上四種指標，算是最常見的監控指標之一，由於本書使用分散式架構
（業界基本上都是分散式），架設了多台雲端機器，那要如何統一監控，
多台雲端機器的系統狀態呢？這就是本章節的重點之一。

14.2.1 Promethus 介紹

首先，什麼是 Prometheus 呢？ Prometheus 是一個開源的系統監控工具、
資料庫，擁有資料收集、儲存、查詢等功能，且具有以下幾項特點：

1. 時間序列資料庫。由於監控系統，大多都是時間序列資料，因此
   Prometheus 最常與 Grafana 一起使用。
2. 與多種服務高度相容。如果單純只有時間序列資料庫的特性，不足以
   與 Grafana 同時使用，因此，Prometheus 第二個特色，與現代化各種
   服務監控相容，如 MySQL、Netdata、Redis、Cadvisor、Traefik 等
   等，甚至官方有推出 Python Package - prometheus-client，可用 Python
   客製化監控指標。
3. 擁有豐富的插鍵。即使官方沒有原生支援，Prometheus 社群也擁有大
   量的第三方工具，提供監控使用。
4. 視覺化工具。Prometheus 原生就擁有視覺化版面，但功能相對基礎，
   大多時候會與專門的視覺化工具 - Grafana 搭配使用。

單純看文字，可能不容易理解，以下將直接架設 Prometheus，讓讀者更
有感：

使用以下 yml 檔，即可架設。

https://github.com/FinMind/FinMindBook/blob/master/DataEngineering/
Chapter14/14.2/14.2.1/prometheus.yml

```
version: '3.0'

services:
  prometheus:
    image: prom/prometheus:v2.1.0
    volumes:
      - prometheus_data:/prometheus
    ports:
      - target: 9090
        published: 9090
        mode: host
    user: root
    restart: always
    networks:
        - my_network

networks:
  my_network:

volumes:
  prometheus_data: {}
```

yml 檔細節，就不多作介紹，本書到此，讀者應該都能看懂。

使用以下指令啟動 Prometheus

```
docker-compose -f prometheus.yml up
```

完成後，跳轉到 http://localhost:9090/graph，會出現以下畫面，代表成功架設 Prometheus。

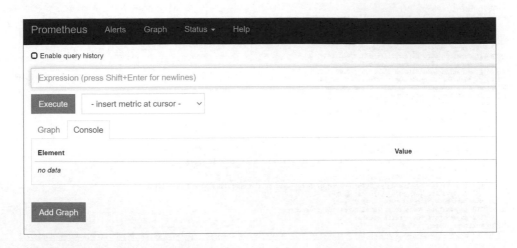

其中，Status 底下的 Targets，代表 Prometheus 會到以下端點 Endpoint，收集監控資料。

注意一點，在監控的世界，metrics 是關鍵字，大多數服務都會將監控數據，放到 metrics 路徑下，再由 Prometheus 進行收集。

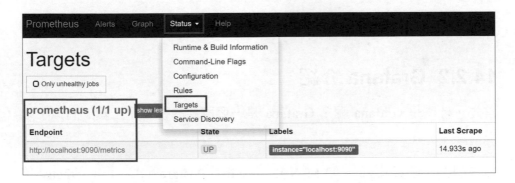

讀者有興趣，可以跳轉網頁到，http://localhost:9090/metrics，會出現以下畫面。

```
←  →  C  ⓘ localhost:9090/metrics

# HELP go_gc_duration_seconds A summary of the GC invocation durations.
# TYPE go_gc_duration_seconds summary
go_gc_duration_seconds{quantile="0"} 0.0001162
go_gc_duration_seconds{quantile="0.25"} 0.000225
go_gc_duration_seconds{quantile="0.5"} 0.0002853
go_gc_duration_seconds{quantile="0.75"} 0.000346
go_gc_duration_seconds{quantile="1"} 0.0005065
go_gc_duration_seconds_sum 0.0039251
go_gc_duration_seconds_count 13
# HELP go_goroutines Number of goroutines that currently exist.
# TYPE go_goroutines gauge
go_goroutines 86
# HELP go_memstats_alloc_bytes Number of bytes allocated and still in use.
# TYPE go_memstats_alloc_bytes gauge
go_memstats_alloc_bytes 2.7690656e+07
# HELP go_memstats_alloc_bytes_total Total number of bytes allocated, even if freed.
# TYPE go_memstats_alloc_bytes_total counter
go_memstats_alloc_bytes_total 7.572356e+07
# HELP go_memstats_buck_hash_sys_bytes Number of bytes used by the profiling bucket hash table.
# TYPE go_memstats_buck_hash_sys_bytes gauge
go_memstats_buck_hash_sys_bytes 1.459215e+06
# HELP go_memstats_frees_total Total number of frees.
# TYPE go_memstats_frees_total counter
go_memstats_frees_total 404847
# HELP go_memstats_gc_cpu_fraction The fraction of this program's available CPU time used by the GC since the program started.
```

這就是 Prometheus 格式。

以上是 Prometheus 主要的架構，後續章節會陸續增加 MySQL、Rabbitmq、Netdata 等端點 Endpoint 到 Prometheus，讓它收集各種服務的監控數據，格式也會與上圖類似。

介紹完 Prometheus 後，接著下一小節，介紹視覺化工具 - Grafana。

14.2.2　Grafana 介紹

首先，什麼是 Grafana 呢？ Grafana 是一個開源的視覺化工具，有以下幾項特點：

1. 支援多種資料庫，如 MySQL、Redis、PostgreSQL、Prometheus 等等，Grafana 使用以上資料庫，進行視覺化監控，而 Prometheus 就是本書使用的資料庫。
2. 強大的視覺化工具，支援折線圖、圓餅圖、表格等等，且單純透過拖拉方式即可製作儀錶板，非常簡單。

3. 可使用各種 SQL 語法進行查詢。如第一點所說，Grafana 支援多種資料庫，且可使用該資料庫的語法，進行查詢、監控，如 SQL 語法、Redis 指令等。

4. 內建警報系統。監控系統，不可能人 24 小時觀察，因此一定需要警報系統，Grafana 支援 Email、Slack、Telegram、Line 等常見的通訊工具，發送警報通知，而觸發條件可自行設定。

5. 豐富的插鍵。與 Prometheus 一樣，Grafana 擁有強大的社群，因此也有豐富的第三方工具可使用。

介紹完特色後，本節一樣直接進行架設，讓讀者更容易了解。

使用以下 yml 檔，即可架設。

https://github.com/FinMind/FinMindBook/blob/master/DataEngineering/Chapter14/14.2/14.2.2/grafana_prometheus.yml

```
version: '3.0'

services:
  prometheus:
    image: prom/prometheus:v2.1.0
    volumes:
      - prometheus_data:/prometheus
    ports:
      - target: 9090
        published: 9090
        mode: host
    user: root
    restart: always
    networks:
      - my_network
```

```
    grafana:
      image: grafana/grafana:8.5.25
      ports:
        - target: 3000
          published: 3000
          mode: host
      user: root
      environment:
        # 設定密碼
        - GF_SECURITY_ADMIN_PASSWORD=pass
      restart: always
      networks:
          - my_network

networks:
  my_network:

volumes:
  prometheus_data: {}
```

使用以下指令啟動 Grafana & Prometheus

```
docker-compose -f grafana_prometheus.yml up
```

架設成功後，跳轉到 Grafana 頁面 - http://localhost:3000。

帳號密碼是，admin / pass。

由於書本印刷，筆者在此先將畫面改成亮色系，讀者可自行根據以下步驟調整：

14-11

接著,新增 Prometheus 作為 Data Source。

設定 Data Source 路徑,因為 Prometheus 與 Grafana 在同一個 Docker Network 底下,因此 URL 是 prometheus:9090。

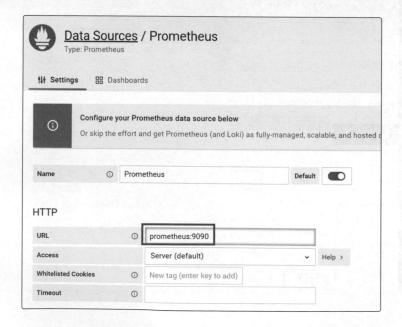

接著到最下面，按 Save & test，儲存設定，並進行測試，測試成功後，
匯出圖 3，代表連線成功。

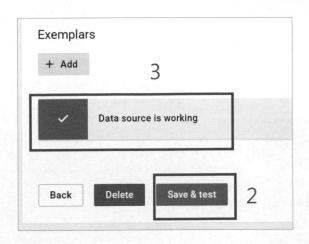

到此步驟後，本書成功架設 Prometheus、Grafana，接著將結合其他服
務，在同一個頁面，監控多個服務。

14.2.3 Netdata 介紹

在連結其他服務之前，先介紹一個，系統監控工具 - Netdata，主要用於
監控 CPU、記憶體 Memory 用量、硬碟 Disk 用量、Network 網路流量、
系統程序等各項指標。

為什麼要介紹 Netdata ？由於本書使用分散式架構，使用多台雲端機器，
因此需要監控機器的 """ 狀態 """，以確保雲端機器負荷正常。

使用以下 yml 檔架設 Netdata。

https://github.com/FinMind/FinMindBook/blob/master/DataEngineering/
Chapter14/14.2/14.2.3/netdata.yml

```
version: '3.0'

services:
  netdata:
    restart: always
    hostname: "netdata"
    ports:
      - 19999:19999
    image: netdata/netdata:v1.31.0
    cap_add:
      - SYS_PTRACE
    security_opt:
      - apparmor:unconfined
    volumes:
      - /proc:/host/proc:ro
      - /sys:/host/sys:ro
      - /var/run/docker.sock:/var/run/docker.sock:ro
    networks:
      - my_network

networks:
  my_network:
```

使用以下指令啟動 Netdata

```
docker-compose -f netdata.yml up
```

成功後,跳轉到以下連結,http://localhost:19999/,會出現以下畫面(讀者看到的應該是暗色系畫面,由於書本,筆者使用亮色系風格)。

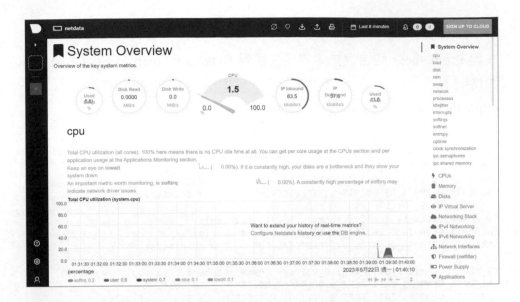

可以看到，Netdata 監控系統各種資訊，如 CPU、Disk 硬碟讀寫。

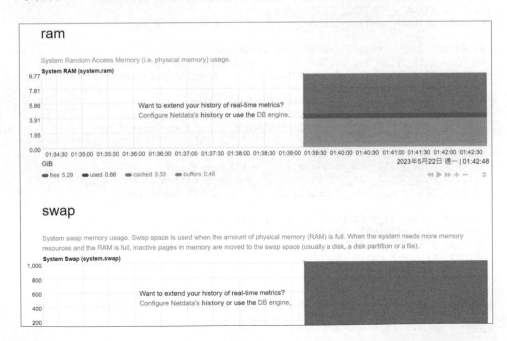

系統記憶體 ram、swap 用量等資訊。

在與 Prometheus 結合之前，需要先找到 Netdata 的 metrics 路徑，請讀者
跳轉到：

http://localhost:19999/api/v1/allmetrics

```
# chart: netdata.db_points (name: netdata.db_points)
NETDATA_NETDATA_DB_POINTS_READ="0"          # points/s
NETDATA_NETDATA_DB_POINTS_GENERATED="-0"        # points/s
NETDATA_NETDATA_DB_POINTS_VISIBLETOTAL="0"        # points/s

# chart: netdata.queries (name: netdata.queries)
NETDATA_NETDATA_QUERIES_QUERIES="0"          # queries/s
NETDATA_NETDATA_QUERIES_VISIBLETOTAL="0"        # queries/s

# chart: sensors.BAT1-acpi-0_voltage (name: sensors.BAT1_acpi_0_voltage)
NETDATA_SENSORS_BAT1_ACPI_0_VOLTAGE_IN0="5"        # Volts
NETDATA_SENSORS_BAT1_ACPI_0_VOLTAGE_VISIBLETOTAL="5"        # Volts

# chart: netdata.runtime_sensors (name: netdata.runtime_sensors)
NETDATA_NETDATA_RUNTIME_SENSORS_RUN_TIME="0"        # ms
NETDATA_NETDATA_RUNTIME_SENSORS_VISIBLETOTAL="0"        # ms

# chart: disk_svctm.sr0 (name: disk_svctm.sr0)
NETDATA_DISK_SVCTM_SR0_SVCTM="0"        # milliseconds/operation
NETDATA_DISK_SVCTM_SR0_VISIBLETOTAL="0"        # milliseconds/operation

# chart: disk_ext_avgsz.sr0 (name: disk_ext_avgsz.sr0)
NETDATA_DISK_EXT_AVGSZ_SR0_DISCARDS="0"        # KiB/operation
NETDATA_DISK_EXT_AVGSZ_SR0_VISIBLETOTAL="0"        # KiB/operation

# chart: disk_avgsz.sr0 (name: disk_avgsz.sr0)
NETDATA_DISK_AVGSZ_SR0_READS="0"        # KiB/operation
NETDATA_DISK_AVGSZ_SR0_WRITES="0"        # KiB/operation
NETDATA_DISK_AVGSZ_SR0_VISIBLETOTAL="0"        # KiB/operation

# chart: disk_ext_await.sr0 (name: disk_ext_await.sr0)
NETDATA_DISK_EXT_AWAIT_SR0_DISCARDS="0"        # milliseconds/operation
NETDATA_DISK_EXT_AWAIT_SR0_FLUSHES="0"        # milliseconds/operation
NETDATA_DISK_EXT_AWAIT_SR0_VISIBLETOTAL="0"        # milliseconds/operation
```

以上就是 Netdata 的 metrics，基本上與 Prometheus 格式一致，可以直接
結合。

14.2.4 如何尋找監控路徑 Metrics ？

一定有許多讀者想問，筆者是如何找到 Netdata 的 metrics 路徑呢？基本上可以從官方文件開始尋找，以 Netdata 為例，如下：

https://learn.netdata.cloud/docs/exporting-data/prometheus/#querying-metrics

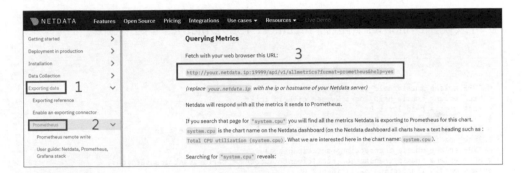

因為本書選用最知名的監控工具，那各種服務一定有支援 Prometheus。

又例如 RabbitMQ 官方文件：

https://www.rabbitmq.com/prometheus.html

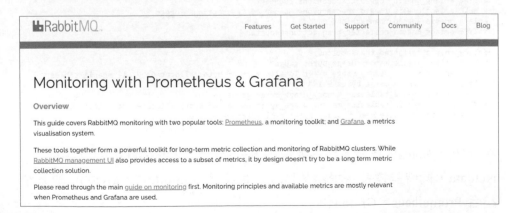

稍微搜尋 metrics 關鍵字後，

To confirm that RabbitMQ now exposes metrics in Prometheus format, get the first couple of lines with `curl` or similar:

```
curl -s localhost:15692/metrics | head -n 3
# TYPE erlang_mnesia_held_locks gauge
# HELP erlang_mnesia_held_locks Number of held locks.
erlang_mnesia_held_locks{node="rabbit@65f1a10aaffa",cluster="rabbit@65f1a10aaffa"} 0
```

Notice that RabbitMQ exposes the metrics on a dedicated TCP port. `15692` by default.

就可以找到，metrics 路徑在 localhost:15692/metrics，如下圖：

```
← → C  ① localhost:15692/metrics

# TYPE erlang_mnesia_held_locks gauge
# HELP erlang_mnesia_held_locks Number of held locks.
erlang_mnesia_held_locks 0
# TYPE erlang_mnesia_lock_queue gauge
# HELP erlang_mnesia_lock_queue Number of transactions waiting for a lock.
erlang_mnesia_lock_queue 0
# TYPE erlang_mnesia_transaction_participants gauge
# HELP erlang_mnesia_transaction_participants Number of participant transactions.
erlang_mnesia_transaction_participants 0
# TYPE erlang_mnesia_transaction_coordinators gauge
# HELP erlang_mnesia_transaction_coordinators Number of coordinator transactions.
erlang_mnesia_transaction_coordinators 0
# TYPE erlang_mnesia_failed_transactions counter
# HELP erlang_mnesia_failed_transactions Number of failed (i.e. aborted) transactions.
erlang_mnesia_failed_transactions 1
# TYPE erlang_mnesia_committed_transactions counter
# HELP erlang_mnesia_committed_transactions Number of committed transactions.
erlang_mnesia_committed_transactions 65
# TYPE erlang_mnesia_logged_transactions counter
# HELP erlang_mnesia_logged_transactions Number of transactions logged.
erlang_mnesia_logged_transactions 81
# TYPE erlang_mnesia_restarted_transactions counter
# HELP erlang_mnesia_restarted_transactions Total number of transaction restarts.
erlang_mnesia_restarted_transactions 0
# TYPE erlang_mnesia_memory_usage_bytes gauge
# HELP erlang_mnesia_memory_usage_bytes Total number of bytes allocated by all mnesia tables
erlang_mnesia_memory_usage_bytes 91144
# TYPE erlang_mnesia_tablewise_memory_usage_bytes gauge
# HELP erlang_mnesia_tablewise_memory_usage_bytes Number of bytes allocated per mnesia table
erlang_mnesia_tablewise_memory_usage_bytes{table="schema"} 29608
```

在介紹完 Netdata 監控系統數據後，下一節將結合 Prometheus、Grafana、Netdata，同時監控多台機器的狀態，最後將 RabbitMQ、MySQL 等系統結合 Prometheus、Grafana。

14.3 架設個人化監控儀表板

介紹完基礎的 Prometheus & Grafana 後，本章節將結合以下服務：

- Netdata：監控分散式下，多台機器效能負荷狀態，避免發生，記憶體用滿、硬碟用滿等狀況。
- MySQL：監控資料庫 """ 負荷 """ 狀態，你不會希望，資料庫負荷太重，影響資料查詢效能。
- RabbitMQ：監控 """ 任務 """ 消化狀態，確保任務都有在時間內被執行。
- Flower：監控 """ 工人 """ 運作狀態，確保工人正常工作。
- Redis：監控負荷狀態。
- Airflow：監控 Airflow 排程執行狀態，確認排程都有正常運作。

在 14.3 節中，本書使用 Docker Swarm 架構，架設監控系統，請讀者先完成以下指令，如已完成 Docker Swarm 可跳過。

切換到以下路徑，執行以下指令：

https://github.com/FinMind/FinMindBook/tree/master/DataEngineering/Chapter14/14.3

Docker Swarm 初始化

```
docker swarm init
```

建立 Portainer

```
docker stack deploy -c portainer.yml
```

建立 docker network

```
docker network create --scope=swarm --driver=overlay my_network
```

完成環境建置後，開始將相關的服務，加入監控。

14.3.1 Netdata

本節將帶著讀者，把 Netdata 與 Prometheus & Grafana 結合。

首先，切換到以下路徑：

https://github.com/FinMind/FinMindBook/tree/master/DataEngineering/
Chapter14/14.3/14.3.1

架構如下：

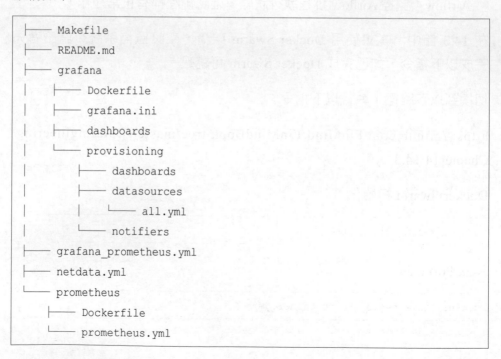

```
├── Makefile
├── README.md
├── grafana
│   ├── Dockerfile
│   ├── grafana.ini
│   ├── dashboards
│   └── provisioning
│       ├── dashboards
│       ├── datasources
│       │   └── all.yml
│       └── notifiers
├── grafana_prometheus.yml
├── netdata.yml
└── prometheus
    ├── Dockerfile
    └── prometheus.yml
```

1. 使用 garfana_prometheus.yml、netdata.yml 架設相關服務。
2. grafana 資料夾底下，是 garfana 相關設定，雖然可以手動調整，但一般來說，都寫進設定檔，並包進 Docker 裡面比較好。
3. prometheus 資料夾底下，是 prometheus 相關設定，一樣將設定檔包進 Docker。

那如何啟動呢？選定一台本地端電腦 or 雲端機器，我們接下來，要部屬監控系統在這台設備上。

第一，先設定 Node 機器的 Labels，如下圖。
（讀者可用自己的電腦架設，或是隨意選一台雲端機器。）

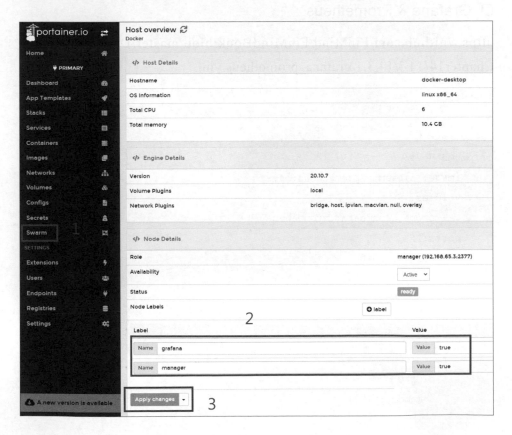

第二，參考以下指令，架設系統。

部屬 Grafana & Prometheus

```
docker stack deploy -c grafana_prometheus.yml monitor
```

部屬 Netdata

```
docker stack deploy -c netdata.yml netdata
```

相關的 yml 檔如下：

❏ Grafana & Prometheus

https://github.com/FinMind/FinMindBook/blob/master/DataEngineering/
Chapter14/14.3/14.3.1/grafana_prometheus.yml

```
version: '3.8'

services:
  prometheus:
    image: linsamtw/prometheus:14.2
    volumes:
      - prometheus_data:/prometheus
    ports:
      - target: 9090
        published: 9090
        mode: host
    user: root
    deploy:
      resources:
        limits:
          memory: 1024M
```

```
      reservations:
        memory: 512M
    replicas: 1
    update_config:
      parallelism: 1
      delay: 5s
      order: stop-first
      failure_action: rollback
    placement:
      constraints: [node.labels.grafana == true]
  restart: always
  networks:
      - my_network

grafana:
  image: linsamtw/grafana:14.2
  ports:
      - target: 3000
      published: 3000
      mode: host
  user: root
  environment:
      - GF_SECURITY_ADMIN_PASSWORD=pass
  deploy:
    replicas: 1
    update_config:
      parallelism: 1
      delay: 10s
    placement:
      constraints: [node.labels.grafana == true]
  restart: always
  networks:
```

```
          - my_network

networks:
  my_network:
    # 加入已經存在的網路
    external: true


volumes:
  prometheus_data: {}
```

❏ Netdata

https://github.com/FinMind/FinMindBook/blob/master/DataEngineering/
Chapter14/14.3/14.3.1/netdata.yml

```
version: '3.7'

# 建立模板
# 由於不同機器，設定都一樣，
# 因此使用模板
x-netdata-service: &netdata-service
    restart: always
    hostname: "{{.Node.Hostname}}-{{.Service.Name}}"
    image: netdata/netdata:v1.31.0
    cap_add:
      - SYS_PTRACE
    security_opt:
      - apparmor:unconfined
    volumes:
      - /proc:/host/proc:ro
      - /sys:/host/sys:ro
      - /var/run/docker.sock:/var/run/docker.sock:ro
```

```
    networks:
      - my_network

services:
  netdata-manager:
    # 引用模板
    <<: *netdata-service
    ports:
      - 19999:19999
    deploy:
      resources:
        limits:
          memory: 128M
        reservations:
          memory: 64M
      placement:
        constraints:
          # 設定，部屬在 manager = true 的機器
          - node.labels.manager == true

networks:
  my_network:
    external: true
```

架設完成後，切換到：

http://localhost:9000/#/services

在 Services 底下，出現三個服務，代表成功部屬 Grafana & Prometheus & Netdata。

接著跳轉到 Prometheus 頁面，http://localhost:9090/targets。

在 Targets 底下，會出現 netdata-manager，代表現在，Prometheus 可以讀取到 Netdata 的監控資訊。

（在此主要是透過，將設定檔包進 Docker Image 中，讓 Prometheus 啟動後，透過設定檔，直接完成 Targets 設定。）

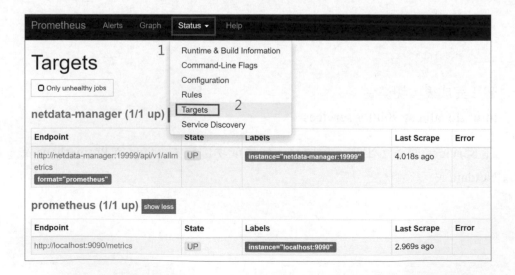

那如何撰寫 Prometheus 設定檔呢？

參考以下 yml 檔。

https://github.com/FinMind/FinMindBook/blob/master/DataEngineering/
Chapter14/14.3/14.3.1/prometheus/prometheus.yml

```
global:
  scrape_interval: 1s
  scrape_timeout: 1s
# 設定 prometheus 監控目標 Targets
scrape_configs:
# 第一個 Targets
# 監控 prometheus 自己本身
- job_name: prometheus
  scrape_interval: 5s
  scrape_timeout: 3s
  metrics_path: /metrics
  static_configs:
  - targets:
    - localhost:9090

# 第二個 Targets - Netdata
# 如讀者有兩台以上的機器
# 可複製以下設定
# 並將 netdata-manager 改成 netdata-worker1...
- job_name: 'netdata-manager'
  scrape_interval: 5s
  scrape_timeout: 5s
  # metrics 路徑
  metrics_path: /api/v1/allmetrics
  params:
```

```
   # metrics 格式
   format: [ prometheus ]
static_configs:
   # Targets 路徑
   - targets: ['netdata-manager:19999']
```

global 底下：

1. scrape_interval：設定抓取資料的間隔時間，本書中設定，一秒抓取一次。

2. scrape_timeout：設定抓取資料的超時時間，本書設定，超過一秒即為超時。

（本書中，想要收集每秒系統狀態，因此都設定 1 秒。）

接著 scrape_configs 底下，主要是用於，設定 Targets 目標。

❑ 第一個 Targets

```
- job_name: prometheus
  scrape_interval: 5s
  scrape_timeout: 3s
  metrics_path: /metrics
  static_configs:
  - targets:
    - localhost:9090
```

1. job_name，代表這個監控目標的名稱，讀者可自行設定，為了方便閱讀，基本上都會設定監控的 service 服務名稱。

2. scrape_interval：設定 5 秒收集一次監控數據。

3. scrape_timeout：設定收集資料超過 3 秒，則超時。

4. metrics_path：監控指標的路徑，此 Targets 是 Prometheus，監控路徑是，http://localhost:9090/metrics，因此 metrics_path 就設定 /metrics。

5. static_configs：底下的 targets，代表監控目標的地址 host 和端口 port。

❑ 第二個 Targets

```
- job_name: 'netdata-manager'
  scrape_interval: 5s
  scrape_timeout: 5s
  # metrics 路徑
  metrics_path: /api/v1/allmetrics
  params:
    # metrics 格式
    format: [ prometheus ]
  static_configs:
    # Targets 路徑
    - targets: ['netdata-manager:19999']
```

第二個 Targets，是 netdata 的監控：

1. metrics_path：監控指標的路徑，在 netdata 中，是 http://localhost:19999/api/v1/allmetrics，可參考 14.2.3。

2. params：設定 metrics 的格式，在此 format 設定為 prometheus。

3. static_configs：因為在 Docker Swarm 網路中，因此 host 是 service 名稱，netdata-manager，port 則是 19999。

以上就是 prometheus 基本設定檔，介紹完後，使用 https://github.com/FinMind/FinMindBook/blob/master/DataEngineering/Chapter14/14.3/14.3.1/prometheus/Dockerfile 這個 Dockerfile，在建立 Docker Image 時，將設定檔加進去。

```
FROM prom/prometheus:v2.1.0

ADD ./prometheus/prometheus.yml /etc/prometheus/
```

以上架設完 Prometheus 後，再跳轉到 grafana 頁面 http://localhost:
3000/，帳號密碼為，admin / pass。

（Grafana 也一樣透過 Dockerfile，https://github.com/FinMind/FinMindBook/
blob/master/DataEngineering/Chapter14/14.3/14.3.1/grafana/Dockerfile，將
設定檔複製進 Docker Image 中。）

由於相關設定，已複製進 Docker 內，因此可以看到，啟動 Grafana 後，
直接變成亮色系主題。

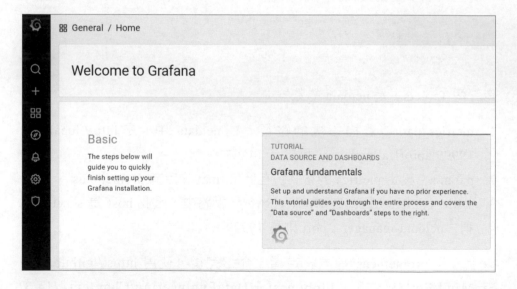

接著到 Configuration 頁面，已經設定好資料源 Data Sources。

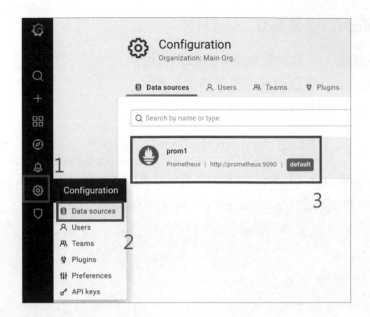

那 Grafana 要如何撰寫設定檔呢？

首先，Grafana 的 Dockerfile 如下：

```
FROM grafana/grafana:8.5.25
COPY ./grafana/provisioning /etc/grafana/provisioning
COPY ./grafana/grafana.ini /etc/grafana/grafana.ini
```

❏ 第一，將 provisioning 複製進 Docker Image 內

```
├── grafana
│   ├── Dockerfile
│   ├── config.ini
│   ├── dashboards
│   └── provisioning
│       ├── dashboards
│       ├── datasources
│       │   └── all.yml
│       └── notifiers
```

provisioning 底下，主要有三個資料夾：

1. dashboards：存放視覺化儀表板設定檔位置。
2. datasources：存放資料源設定檔的位置。
3. notifiers：存放 Alert 警報設定檔位置。

在 14.3.1 章節，先介紹 datasources/all.yml，如下：

https://github.com/FinMind/FinMindBook/blob/master/DataEngineering/
Chapter14/14.3/14.3.1/grafana/provisioning/datasources/all.yml

```
datasources:
- name: 'prom1'
  type: 'prometheus'
  access: 'proxy'
  org_id: 1
  url: 'http://prometheus:9090'
  is_default: true
  version: 1
  editable: true
```

1. name：設定在 Grafana 中，數據源的名稱。
2. type：數據源的類型，此處為 Prometheus。
3. access：設定數據源的訪問方式，此處使用 proxy 代理方式進行訪問。
4. org_id：指定組織的 ID，此處設定為 1。
5. url：指定 Prometheus 的位址，此處設定為 http://prometheus:9090，表示 Grafana 將通過這個 URL 訪問 Prometheus。
6. is_default：設定此數據源是否為默認數據源，此處設定為 true，表示將使用此數據源作為默認。
7. version：指定數據源的版本，此處設定為 1。

8. editable：設定數據源是否可編輯（後續可能做調整），此處設定為 true，表示可以編輯此數據源。

設定完成後，Grafana 在啟動階段，就會套用此設定檔，對相關的資料源做連線。

❏ 第二，將 grafana.ini 複製進 Docker Image 內

可參考以下連結：

https://github.com/FinMind/FinMindBook/blob/master/DataEngineering/
Chapter14/14.3/14.3.1/grafana/grafana.ini

```ini
[paths]
provisioning = /etc/grafana/provisioning

[server]
enable_gzip = true

[users]
default_theme = light
```

1. paths：設定 Grafana 設定文件的路徑位置，在此設定為 /etc/grafana/provisioning。
2. server：設定啟動 enable_gzip = true，啟動壓縮功能，提高效能。
3. users：設定 Grafana 預設風格，由於書本印刷，在此設定亮色系。

介紹完相關設定後，開始製作 Netdata 的儀錶板 Dashboard。

14.3.1.1 製作 Dashboard

下圖是本書使用的 Netdata Dashboard，可監控 CPU 用量、Memory 用量、Disk IO 等系統資訊，那要如何製作呢？

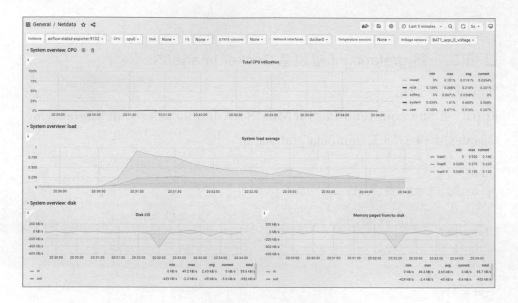

最基本的方式，就是手動從 0 開始一個個圖表製作，但 Grafana 官方有相關的模板，可直接使用，相關連結如下：

https://grafana.com/grafana/dashboards/

到以上連結後，可搜尋相關 Dashboard，在此搜尋 Netdata，會出現多個模板。

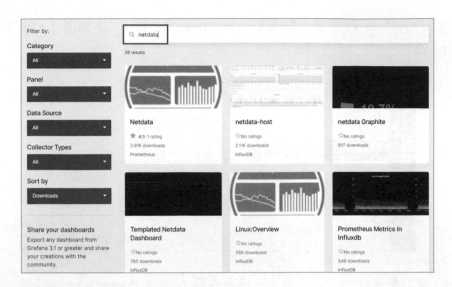

筆者查看幾個模板後,選用以下連結:

https://grafana.com/grafana/dashboards/7107-netdata/

那要如何使用模板呢?如下圖,先點選 Copy ID to clipboard,複製儀錶板 ID。

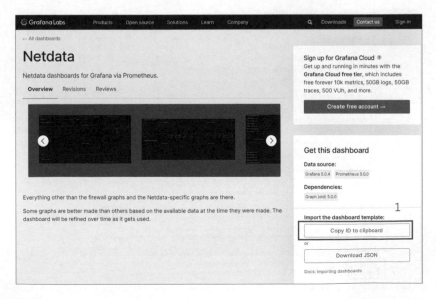

跳轉到 Grafana 頁面後，點選 Create、Import，將 ID 貼到圖 4，接著點選 Load。

Load 讀取後，會出現以下畫面，接著在圖 6 選擇 Prometheus 資料源，最後點選 Import 導入儀錶板。

最後，就完成了儀表板，有非常多圖表可以觀察。

讀者到此，是不是很有成就感呢？

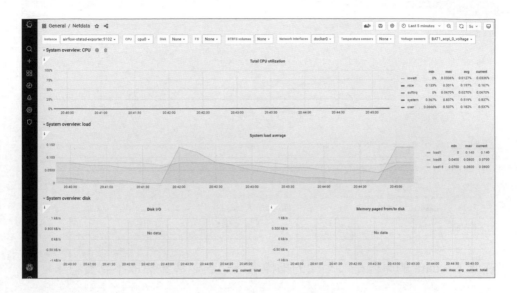

但有些圖表顯示，No data，這又是什麼原因呢？

點選左下角，Disk I/O 的位置，點選 Edit，如下圖：

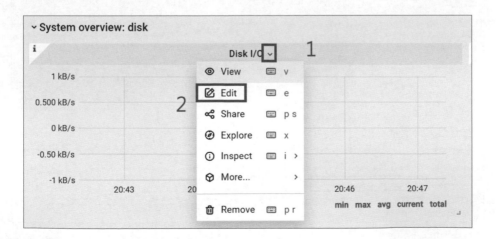

會跳轉到以下，編輯 Chart 畫面。

在此 No data 的原因是，相關的指標名稱錯誤，導致抓取不到資料。

讀者可按照下圖，修改 Metrics 名稱，就會正常顯示圖表了。

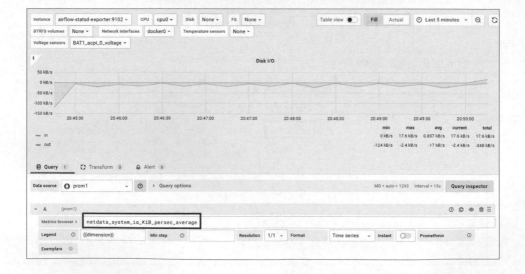

那筆者是如何知道，要輸入什麼關鍵字呢？

如下圖，可以先輸入前面幾個關鍵字後，（如 netdata_system_io），Garfana 會自動帶出，類似的指標提供使用，因此就能找到，正確的指標名稱了。

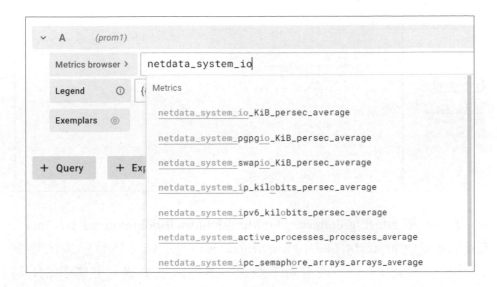

同理，出現 No data 的圖表，可以透過以上方式，做調整。

最後調整完後，點選右上角的齒輪。

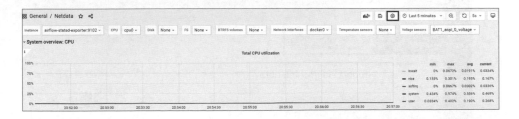

進入 Settings 後，點選 JSON Model，出現相對應的 JSON 格式，這裡告訴讀者，Grafana 的儀表板背後是透過 JSON 做儲存。

讀者可以將相關的 Dashboard 都用 JSON 儲存，紀錄調整過後的儀表板，避免系統出現問題時，相關 Dashboard 遺失。

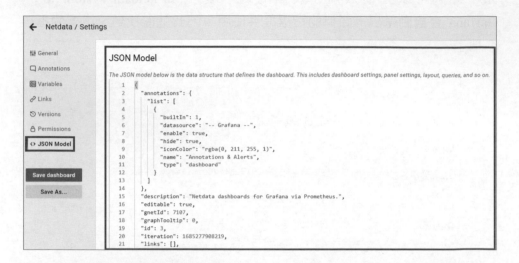

以下連結 https://github.com/FinMind/FinMindBook/blob/master/Data Engineering/Chapter14/14.3/14.3.1/netdata.json，是筆者調整過後的儀表板，並存成 JSON 檔。讀者可下載後，透過以下方式，上傳 JSON 到 Grafana。

圖 4，如果名稱重複，讀者可自行調整，最後在圖 5 點選 Import。

最後可以看到，相關指標都正常了。

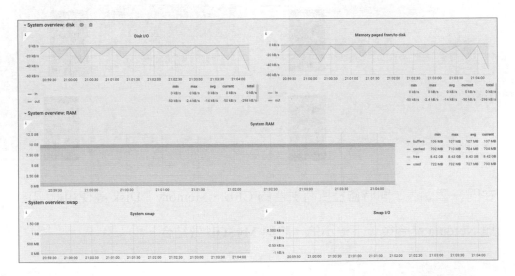

本節，透過 Grafana & Netdata，可觀察雲端 / 本地端設備，相關的系統資訊，如果讀者的系統，有多台設備，則可調整 https://github.com/FinMind/FinMindBook/blob/master/DataEngineering/Chapter14/14.3/14.3.1/netdata.yml，增加多個 netdata services，讓 netdata 部屬在所有設備上。

接著在 https://github.com/FinMind/FinMindBook/blob/master/DataEngineering/Chapter14/14.3/14.3.1/prometheus/prometheus.yml 中，設定多個 job_name，最後，以下是筆者監控 13 台設備的圖表。

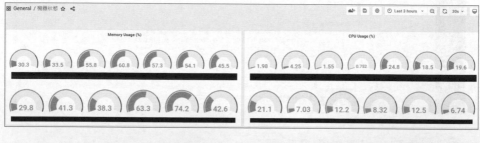

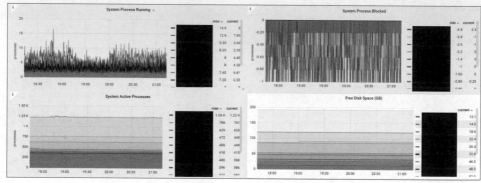

在同一個圖表，同時監控所有機器的 CPU、Memory、Disk 等使用狀況。

與 Netdata 結合後，下一小節，將與 MySQL 做結合。

14.3.2 MySQL

本節將把 MySQL 與 Prometheus & Grafana 結合。

首先，切換到以下路徑：

https://github.com/FinMind/FinMindBook/tree/master/DataEngineering/
Chapter14/14.3/14.3.2

架構如下：

```
├── Makefile
├── README.md
├── grafana
│   ├── Dockerfile
│   ├── dashboards
│   ├── grafana.ini
│   └── provisioning
│       ├── dashboards
│       ├── datasources
│       │   └── all.yml
│       └── notifiers
├── grafana_prometheus.yml
├── mysql.yml
├── netdata.json
├── netdata.yml
└── prometheus
    ├── Dockerfile
    └── prometheus.yml
```

與 14.3.1 類似架構，新增了 mysql.yml，並且在 prometheus.yml 做了一些
調整，如下。

❑ mysql.yml

https://github.com/FinMind/FinMindBook/blob/master/DataEngineering/
Chapter14/14.3/14.3.2/mysql.yml

```yaml
version: '3'
services:

  mysql:
    image: mysql:8.0
    command: mysqld --default-authentication-plugin=mysql_native_password
    ports:
      - 3306:3306
    environment:
      MYSQL_DATABASE: financialdata
      MYSQL_USER: user
      MYSQL_PASSWORD: test
      MYSQL_ROOT_PASSWORD: test
    volumes:
      - mysql:/var/lib/mysql
    # swarm 設定
    deploy:
      mode: replicated
      replicas: 1
      placement:
        constraints: [node.labels.mysql == true]
    networks:
      - my_network

  mysql-exporter:
    image: prom/mysqld-exporter:v0.14.0
    ports:
      - "9104:9104"
```

```
    environment:
      - DATA_SOURCE_NAME=root:test@(mysql:3306)/financialdata
      # swarm 設定
    deploy:
      mode: replicated
      replicas: 1
      placement:
        constraints: [node.labels.mysql == true]
    networks:
        - my_network


phpmyadmin:
    image: phpmyadmin/phpmyadmin:5.1.0
    links:
        - mysql:db
    ports:
        - 8080:80
    environment:
        MYSQL_USER: user
        MYSQL_PASSWORD: test
        MYSQL_ROOT_PASSWORD: test
        PMA_HOST: mysql
    depends_on:
      - mysql
    # swarm 設定
    deploy:
      mode: replicated
      replicas: 1
      placement:
        constraints: [node.labels.mysql == true]
    networks:
        - my_network
```

```
networks:
  my_network:
    # 加入已經存在的網路
    external: true


volumes:
  mysql:
    external: true
```

基本上與 12.5 的 mysql.yml 類似，由於需要與 Prometheus 串接，新增了 mysql-exporter 這個服務，使用 prom/mysqld-exporter:v0.14.0 架設，主要是 Prometheus 推出的一個，用於監控 MySQL 的工具，主要用於收集相關指標，並將資訊導出到 9104 port。

讀者可跳轉到 http://localhost:9104/metrics，查看相關 log 指標，如下圖：

```
# HELP go_gc_cycles_automatic_gc_cycles_total Count of completed GC cycles generated by the Go runtime.
# TYPE go_gc_cycles_automatic_gc_cycles_total counter
go_gc_cycles_automatic_gc_cycles_total 1912
# HELP go_gc_cycles_forced_gc_cycles_total Count of completed GC cycles forced by the application.
# TYPE go_gc_cycles_forced_gc_cycles_total counter
go_gc_cycles_forced_gc_cycles_total 0
# HELP go_gc_cycles_total_gc_cycles_total Count of all completed GC cycles.
# TYPE go_gc_cycles_total_gc_cycles_total counter
go_gc_cycles_total_gc_cycles_total 1912
# HELP go_gc_duration_seconds A summary of the pause duration of garbage collection cycles.
# TYPE go_gc_duration_seconds summary
go_gc_duration_seconds{quantile="0"} 2.38e-05
go_gc_duration_seconds{quantile="0.25"} 0.0001878
go_gc_duration_seconds{quantile="0.5"} 0.0002719
go_gc_duration_seconds{quantile="0.75"} 0.0003681
go_gc_duration_seconds{quantile="1"} 0.0010046
go_gc_duration_seconds_sum 0.5016797
go_gc_duration_seconds_count 1912
# HELP go_gc_heap_allocs_by_size_bytes_total Distribution of heap allocations by approximate size. Note
that this does not include tiny objects as defined by /gc/heap/tiny/allocs:objects, only tiny blocks.
```

接著介紹下一個 yml 檔。

❑ prometheus.yml

在新增 mysql 服務後，需要讓 Prometheus 的 Targets，也新增一個設定，讓 Prometheus 能讀取 mysql 的指標，yml 檔如下：

https://github.com/FinMind/FinMindBook/blob/master/DataEngineering/Chapter14/14.3/14.3.2/prometheus/prometheus.yml

```
global:
# … 略過
scrape_configs:
# … 略過

# 監控 mysql
- job_name: 'mysql'
  scrape_interval: 5s
  scrape_timeout: 5s
  metrics_path: /metrics
  static_configs:
    - targets: ['mysql-exporter:9104']
```

新增一個 mysql：

1. metrics_path：設定路徑在 /metrics。
2. static_configs：由於在同一個 Docker Network 底下，因此設定目標路徑在 mysql-exporter:9104。

只需調整以上兩個設定即可，接下來將架設 MySQL。

（在此之前，需要先在 Label，增加 Name = mysql，Value = true。）

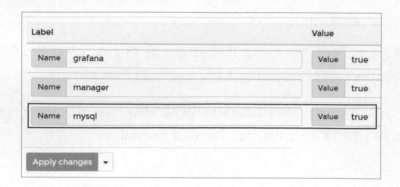

部屬 MySQL

```
docker stack deploy -c mysql.yml mysql
```

更新 Prometheus & Grafana

```
docker stack deploy -c grafana_prometheus.yml monitor
```

完成後，Portainer 就會出現一個，mysql_mysql-exporter 的 services 服務、且 prometheus、grafana 也更新了版本到本節 14.3.2 版本。

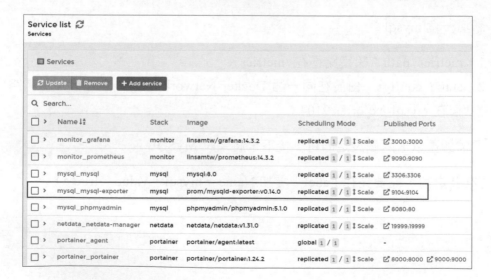

之後跳轉到 Prometheus 頁面，http://localhost:9090/targets，可以看到，
在 Targets 出現 mysql ，代表 Prometheus 會根據這個目標位置，收集
MySQL 指標。

收集完資料後，到 Grafana 增加相對應的儀表板。

14.3.2.1 製作 Dashboard

還記得在 14.3.1.1，建立 Dashboard 時，直接引用 Grafana 模板嗎？本
書也幫讀者挑選好模板了，引用 https://grafana.com/grafana/dashboards/
7362-mysql-overview/ 模板建立 Dashboard。

請讀者先按照下圖，複製 Dashboard ID。

接著按照下圖方式，引入 MySQL 監控模板。

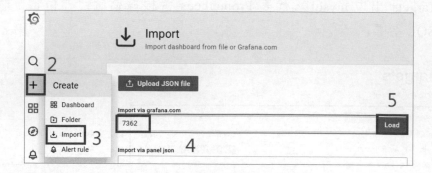

最後選擇 Prometheus 資料源，點選 Import。

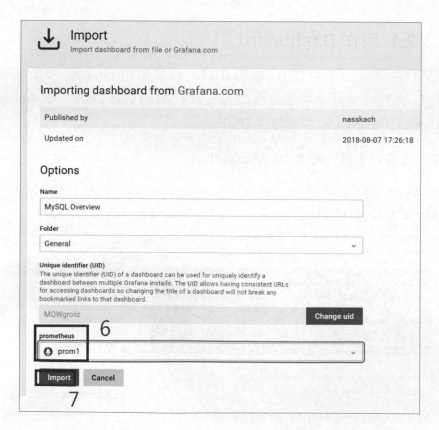

引入模板後，出現以下畫面：

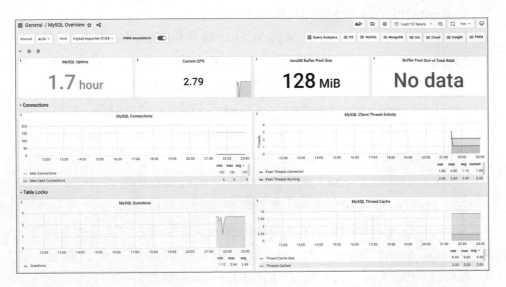

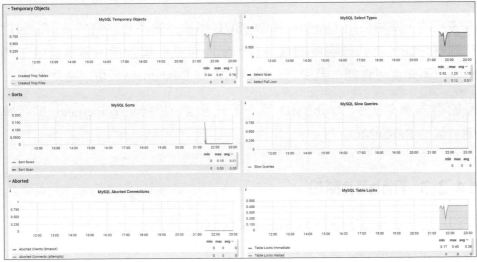

讀者可能會覺得很複雜，如何解讀這麼多圖表，本書將介紹每個 Chart 意義：

1. MySQL Uptime：MySQL 運作時間。
2. Current QPS：當前 MySQL 每秒查詢數量。
3. InnoDB Buffer Pool Size：MySQL 中，緩存數據占用的大小。
4. MySQL Connections：代表當前 MySQL 的連線數量。通常來說，預設值是 100，所以如果此數值超過 100，可能會發生斷線等訊息。
5. MySQL Client Thread Activity：Client 客戶端的活動狀態。
6. MySQL Quertions：表時從 MySQL 收到的查詢數量。這個指標可以用來衡量數據庫的工作負載和查詢的頻率。
7. MySQL Thread Cache：這個指標顯示了線程緩存的使用情況。一般來說，MySQL 使用線程緩存，管理客戶端的分配與回收。
8. MySQL Temporary Objects：臨時物件，包含臨時表、臨時的查詢等。
9. MySQL Select Types：表示 MySQL 中，不同類型的查詢數量。如 Select、Full Join、Left Join、Right Join 等各種查詢。

本書到此，成功使用 Grafana 與 MySQL 串接，是不是非常棒？對於監控初學者來說，有一個現成的儀錶板，幫助監控各種指標，當系統發生問題時，不知從何查起，至少有一個最基本的 Dashboard，可以輔助查看。

還記得本書架設的 RabbitMQ 嗎？下一小節，將與 Prometheus 進行串接，監控 RabbitMQ。

14.3.3 RabbitMQ

本節將把 RabbitMQ 與 Prometheus & Grafana 結合。

首先，切換到以下路徑：

https://github.com/FinMind/FinMindBook/tree/master/DataEngineering/Chapter14/14.3/14.3.3

架構如下：

```
├── Makefile
├── README.md
├── crawler.yml
├── grafana
│   ├── Dockerfile
│   ├── dashboards
│   ├── grafana.ini
│   └── provisioning
│       ├── dashboards
│       ├── datasources
│       │   └── all.yml
│       └── notifiers
├── grafana_prometheus.yml
├── mysql.yml
├── netdata.json
├── netdata.yml
├── prometheus
│   ├── Dockerfile
│   └── prometheus.yml
├── rabbitmq
│   └── Dockerfile
├── rabbitmq.json
└── rabbitmq.yml
```

與前兩節的架構都一樣，本節新增了 rabbitmq.yml、rabbitmq 資料夾、
crawler.yml，同樣在 prometheus.yml 做了一些調整，如下。

❑ rabbitmq.yml

https://github.com/FinMind/FinMindBook/blob/master/DataEngineering/
Chapter14/14.3/14.3.3/rabbitmq.yml

```yaml
version: '3'
services:

  rabbitmq:
    image: 'linsamtw/rabbitmq:14.3.3'
    ports:
      - '5672:5672'
      - '15672:15672'
      # 新增 metrics 的 port，用於監控
      - '15692:15692'
    environment:
      RABBITMQ_DEFAULT_USER: "worker"
      RABBITMQ_DEFAULT_PASS: "worker"
      RABBITMQ_DEFAULT_VHOST: "/"
    # swarm 設定
    deploy:
      mode: replicated
      replicas: 1
      placement:
        constraints: [node.labels.rabbitmq == true]
    networks:
      - my_network

  flower:
    image: mher/flower:0.9.5
    command: ["flower", "--broker=amqp://worker:worker@rabbitmq",
"--port=5555"]
    ports:
      - 5555:5555
    depends_on:
      - rabbitmq
    # swarm 設定
```

```
   deploy:
     mode: replicated
     replicas: 1
     placement:
       constraints: [node.labels.flower == true]
   networks:
     - my_network


networks:
  my_network:
    # 加入已經存在的網路
    external: true
```

增加了 15692 port，用於 metrics 的監控。除此之外，需要啟動 RabbitMQ 的插鍵，才能與 Prometheus 連結，方式如下：

❑ Dockerfile

在原始 RabbitMQ 的 Docker Image 下：
https://github.com/FinMind/FinMindBook/blob/master/DataEngineering/
Chapter14/14.3/14.3.3/rabbitmq/Dockerfile

```
FROM rabbitmq:3.8-management


RUN rabbitmq-plugins enable rabbitmq_prometheus
```

增加一行指令，rabbitmq-plugins enable rabbitmq_prometheus，用於啟動 rabbitmq 連結 prometheus 功能。

❑ prometheus.yml

接著在 prometheus.yml，加入 RabbitMQ metrics 的設定，如下：

https://github.com/FinMind/FinMindBook/blob/master/DataEngineering/
Chapter14/14.3/14.3.3/prometheus/prometheus.yml

```
global:
# … 略過
scrape_configs:
# … 略過
# 監控 rabbitmq
- job_name: 'rabbitmq'
  scrape_interval: 5s
  scrape_timeout: 5s
  metrics_path: /metrics
  static_configs:
    - targets: ['rabbitmq:15692']
```

要模擬監控 RabbitMQ，就需要兩個工具，來達到真實情境，那就是架設 Worker 工人、並且發送任務，本書也準備好了，如下（引用 7.2.1 的 yml 檔）：

❑ crawler.yml

https://github.com/FinMind/FinMindBook/blob/master/DataEngineering/
Chapter14/14.3/14.3.3/crawler.yml

```
version: '3.8'
services:
  crawler_twse:
    image: linsamtw/crawler:7.2.1
    hostname: "twse"
    command: pipenv run celery -A financialdata.tasks.worker worker
--loglevel=info --concurrency=1  --hostname=%h -Q twse
    restart: always
    # swarm 設定
```

```
    deploy:
      mode: replicated
      replicas: 1
      placement:
        constraints: [node.labels.crawler_twse == true]
    environment:
      - TZ=Asia/Taipei
    networks:
        - my_network

  crawler_tpex:
    image: linsamtw/crawler:7.2.1
    hostname: "tpex"
    command: pipenv run celery -A financialdata.tasks.worker worker
--loglevel=info --concurrency=1  --hostname=%h -Q tpex
    restart: always
    # swarm 設定
    deploy:
      mode: replicated
      replicas: 1
      placement:
        constraints: [node.labels.crawler_tpex == true]
    environment:
      - TZ=Asia/Taipei
    networks:
        - my_network

  sent_task:
    image: linsamtw/crawler:7.2.1
    hostname: "sent_task"
    command: pipenv run python financialdata/producer.py taiwan_stock_
price 2021-04-01 2023-04-12
    restart: on-failure
    # swarm 設定
```

```
    deploy:
      mode: replicated
      replicas: 1
      placement:
        constraints: [node.labels.sent_task == true]
    environment:
      - TZ=Asia/Taipei
    networks:
        - my_network


networks:
  my_network:
    # 加入已經存在的網路
    external: true
```

建立兩個 Worker 進行模擬，並且新增 services - sent_task，模擬發送任務情境。

接下來，請讀者按照以下指令，架設 RabbitMQ、Worker、更新 Prometheus、Grafana。

（在此之前，需要先在 Label，增加 crawler_tpex、crawler_twse、flower、rabbitmq、sent_task。）

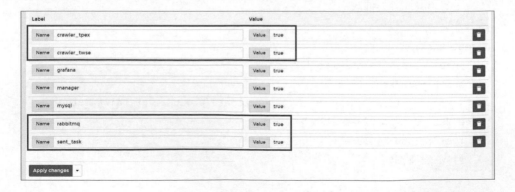

架設 RabbitMQ

```
docker stack deploy -c rabbitmq.yml rabbitmq
```

建立 Worker

```
docker stack deploy -c crawler.yml crawler
```

更新 Prometheus、Grafana

```
docker stack deploy -c grafana_prometheus.yml monitor
```

完成後，Portainer 就會 rabbitmq、flower、crawler、sent_task。

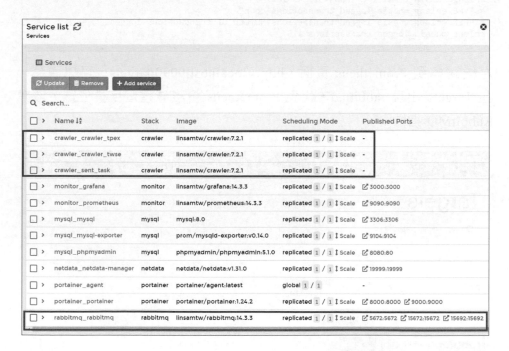

部屬完成後，讀者可以跳轉到 http://localhost:15692/metrics 頁面，可以看到 Rabbitmq 的 metrics，就是 Prometheus 的格式了，後續 Prometheus 就

可以收集 metrics 指標等資訊，如下圖。

```
# TYPE erlang_mnesia_held_locks gauge
# HELP erlang_mnesia_held_locks Number of held locks.
erlang_mnesia_held_locks 0
# TYPE erlang_mnesia_lock_queue gauge
# HELP erlang_mnesia_lock_queue Number of transactions waiting for a lock.
erlang_mnesia_lock_queue 0
# TYPE erlang_mnesia_transaction_participants gauge
# HELP erlang_mnesia_transaction_participants Number of participant transactions.
erlang_mnesia_transaction_participants 0
# TYPE erlang_mnesia_transaction_coordinators gauge
# HELP erlang_mnesia_transaction_coordinators Number of coordinator transactions.
erlang_mnesia_transaction_coordinators 0
# TYPE erlang_mnesia_failed_transactions counter
# HELP erlang_mnesia_failed_transactions Number of failed (i.e. aborted) transactions.
erlang_mnesia_failed_transactions 1
# TYPE erlang_mnesia_committed_transactions counter
# HELP erlang_mnesia_committed_transactions Number of committed transactions.
erlang_mnesia_committed_transactions 161
# TYPE erlang_mnesia_logged_transactions counter
# HELP erlang_mnesia_logged_transactions Number of transactions logged.
erlang_mnesia_logged_transactions 115
```

之後跳轉到 Prometheus 頁面，http://localhost:9090/targets，可以看到，在 Targets 出現 rabbitmq，代表 Prometheus 會根據這個目標位置，收集 RabbitMQ 指標。

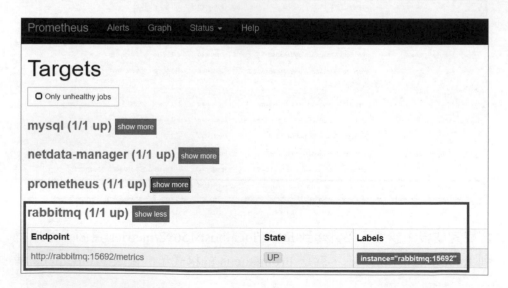

同樣，設定完相關服務後，開始製作 RabbitMQ 的 Dashboard。

14.3.3.1 製作 Dashboard

本書也幫讀者挑選好模板了，引用 https://grafana.com/grafana/dashboards/
4371-rabbitmq-metrics/ 模板建立 Dashboard。

一樣，請讀者先按照下圖，複製 Dashboard ID。

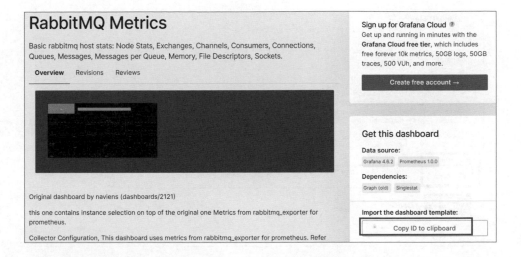

接著按照下圖方式，引入 RabbitMQ 監控模板。

最後選擇 Prometheus 資料源，點選 Import。

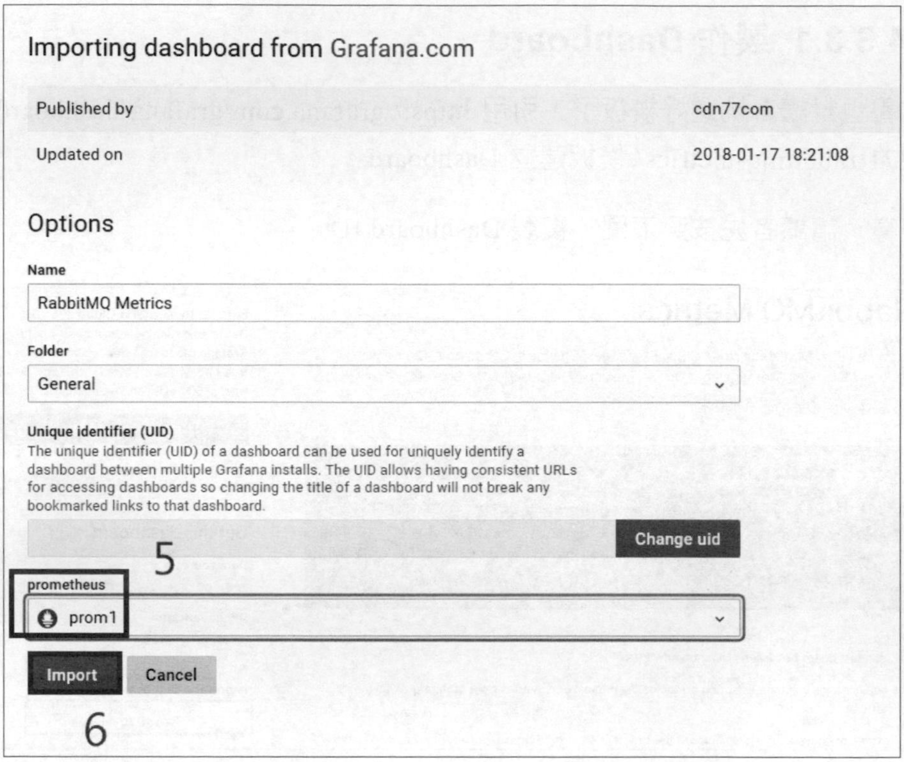

Importing dashboard from Grafana.com

Published by	cdn77com
Updated on	2018-01-17 18:21:08

Options

Name

RabbitMQ Metrics

Folder

General

Unique identifier (UID)
The unique identifier (UID) of a dashboard can be used for uniquely identify a dashboard between multiple Grafana installs. The UID allows having consistent URLs for accessing dashboards so changing the title of a dashboard will not break any bookmarked links to that dashboard.

Change uid

5

prometheus

prom1

Import　　Cancel

6

引入模板後，出現以下畫面。

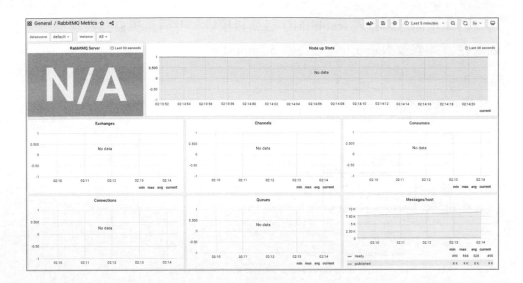

出現很多 No data，這時的狀況，原因與 14.3.1.1 一樣，一些關鍵字稍微出現變動，本書在此，也幫讀者準備好 rabbitmq.json 了。

下載以下連結：

https://github.com/FinMind/FinMindBook/blob/master/DataEngineering/Chapter14/14.3/14.3.3/rabbitmq.json

建立 Dashboard，上傳 JSON 檔。

這時因為前面已經導入 RabbitMQ 儀錶板模板，名稱重複會有錯誤訊息，因此在圖 4 修改 Name，最後在圖 5 Import。

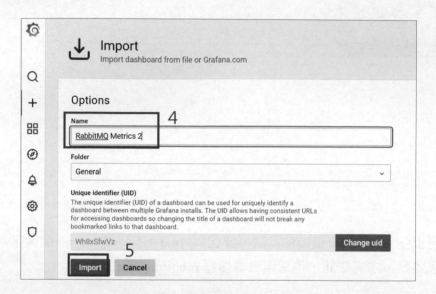

完成後，如下圖。

本書同樣介紹以上圖表的意義：

1. RabbitMQ Server：服務運作了多久時間。
2. Channels：表示當前運作的 Channel 數量。RabbitMQ 透過獨立的 Channel 做訊息隔離。
3. Consumers：表示當前的消費者數量，可以想像成有多少個活躍工人的數量。
4. Connections：表示當前與 RabbitMQ 連接的數量。
5. Queues：表示當前有多少個任務，正在排隊，等待被執行。
6. Messages / host：表示收到任務訊息的總數量。
7. Message / Queue：表示平均每個 Queue 列隊的任務數量。
8. Memory：表示 RabbitMQ 的記憶體 s 用量，如果用量太高就要注意了。
9. Sockets：Sockets 連線數量。

到此，已經完成了 Netdata 監控設備、MySQL 負載狀態監控、RabbitMQ 任務消化狀況監控，接下來將監控 Airflow。

14.3.4 Airflow

本節將把 Airflow 與 Prometheus & Grafana 結合，並引用「12.7 - 常見 Operator 介紹」與「12.8 - Airflow 結合爬蟲 - CeleryExecutor」，架構會稍微複雜，但更接近業界。在真實世界，大多使用分散式的 Airflow 架構，因此本書盡可能模擬真實情境。

首先，切換到以下路徑：

https://github.com/FinMind/FinMindBook/tree/master/DataEngineering/Chapter14/14.3/14.3.4

架構如下：

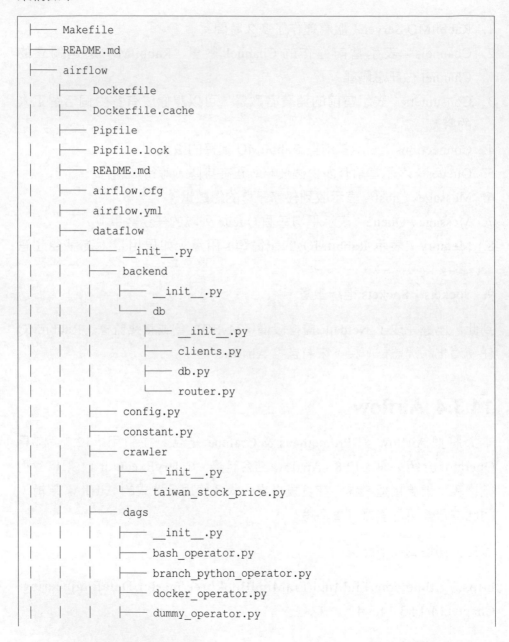

```
├── Makefile
├── README.md
├── airflow
│   ├── Dockerfile
│   ├── Dockerfile.cache
│   ├── Pipfile
│   ├── Pipfile.lock
│   ├── README.md
│   ├── airflow.cfg
│   ├── airflow.yml
│   ├── dataflow
│   │   ├── __init__.py
│   │   ├── backend
│   │   │   ├── __init__.py
│   │   │   └── db
│   │   │       ├── __init__.py
│   │   │       ├── clients.py
│   │   │       ├── db.py
│   │   │       └── router.py
│   │   ├── config.py
│   │   ├── constant.py
│   │   ├── crawler
│   │   │   ├── __init__.py
│   │   │   └── taiwan_stock_price.py
│   │   ├── dags
│   │   │   ├── __init__.py
│   │   │   ├── bash_operator.py
│   │   │   ├── branch_python_operator.py
│   │   │   ├── docker_operator.py
│   │   │   ├── dummy_operator.py
```

```
│     │     │     ├──── python_operator.py
│     │     │     └──── taiwan_stock_price.py
│     │     ├──── etl
│     │     │     ├──── __init__.py
│     │     │     ├──── bash_operator.py
│     │     │     ├──── branch_python_operator.py
│     │     │     ├──── docker_operator.py
│     │     │     ├──── dummy_operator.py
│     │     │     ├──── python_operator.py
│     │     │     └──── taiwan_stock_price.py
│     │     └──── schema
│     │           ├──── __init__.py
│     │           └──── dataset.py
│     ├──── genenv.py
│     ├──── local.ini
│     └──── setup.py
├──── airflow.json
├──── grafana
│     ├──── Dockerfile
│     ├──── dashboards
│     ├──── grafana.ini
│     └──── provisioning
│           ├──── dashboards
│           ├──── datasources
│           │     └──── all.yml
│           └──── notifiers
├──── grafana_prometheus.yml
├──── mysql.yml
├──── netdata.json
├──── netdata.yml
├──── prometheus
```

```
|       ├── Dockerfile
|       └── prometheus.yml
├── rabbitmq
|       └── Dockerfile
├── rabbitmq.json
└── rabbitmq.yml
```

是不是感到越來越複雜了呢？基本上從 14.3.1 開始，先建立 Netdata、再到 14.3.2 建立 MySQL，最後 14.3.3 建立 RabbitMQ，融合以上架構後，本節再與 12.7、12.8 的 Airflow 架構結合。

常見的教學，都是各別建立單一系統，而本書融合整個大數據架構、多個系統的監控，讓讀者更接近實務面。但讀者不用擔心，筆者將整個架構，拆成各個小節，每一小節只對小部分做調整，一步步跟著操作，就能學會。

讓筆者來一一介紹本節新增的部分：

1. airflow 資料夾：基本上是複製 12.7 與 12.8 的程式碼，系統相對複雜，才能感受到真實 Airflow 中，多 DAG 的情境。
2. 移除 14.3.3 的 crawler.yml：本節使用 Airflow 爬蟲取代 crawler.yml。
3. grafana/provisioning/datasources/all.yml：新增 Grafana 的 Data Source - MySQL，因此在這增加設定（由於本書 Airflow 使用 MySQL 作為資料庫，因此在監控部分，選擇透過 MySQL 收集 Airflow 相關資訊）。

grafana/provisioning/datasources/all.yml 的改動如下：

https://github.com/FinMind/FinMindBook/blob/master/DataEngineering/Chapter14/14.3/14.3.4/grafana/provisioning/datasources/all.yml

```
apiVersion: 1

datasources:
- name: 'prom1'
  type: 'prometheus'
  access: 'proxy'
  org_id: 1
  url: 'http://prometheus:9090'
  is_default: true
  version: 1
  editable: true

- name: MySQL-Airflow
  type: mysql
  url: mysql:3306
  database: airflow
  user: airflow
  password: your_password
  secureJsonData:
    password: your_password
```

只需增加 MySQL-Airflow 部分的設定檔，就能在啟動 Grafana 後，自動增加 Data Source。

以上，就是本節新增的部分，只有 3 點，是不是相對簡單呢？一步步說明，能讓讀者更容易學習。下一步，就開始架設 Airflow 吧！

（先記得在 Label 的部分，增加 airflow，後續才能在相對應的設備上，進行部屬。）

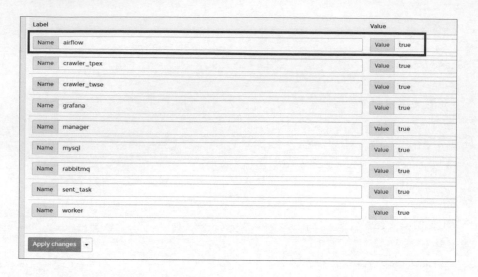

部屬 Airflow

```
docker stack deploy -c airflow/airflow.yml airflow
```

成功後，會出現以下畫面：

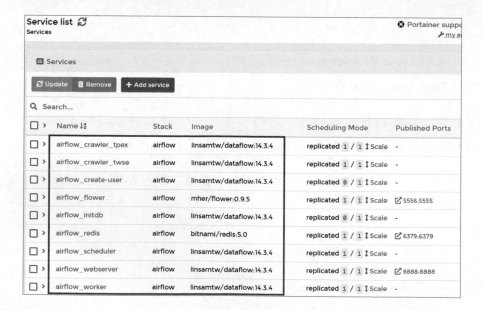

再跳轉到 Airflow 頁面：

http://localhost:8888/home

（如果 DAG 開關是關閉的，再請讀者打開開關。）

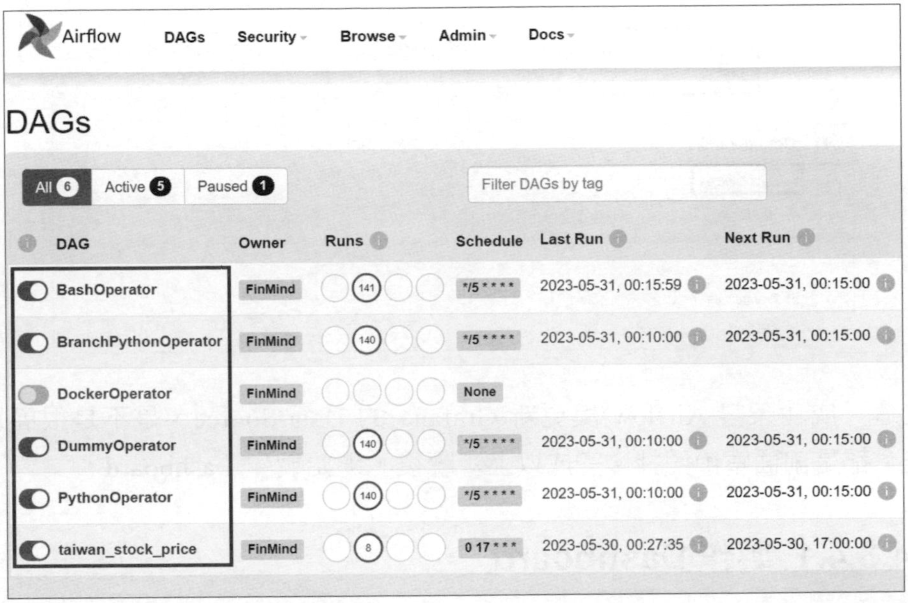

這時如同 12.8，成功部屬 Airflow。

接著再更新 Prometheus & Grafana：

```
docker stack deploy -c grafana_prometheus.yml monitor
```

更新完後，跳轉到 Grafana 頁面：

http://localhost:3000/

按照以下步驟，到 Data Sources 頁面，就會出現 MySQL-Airflow 啦！

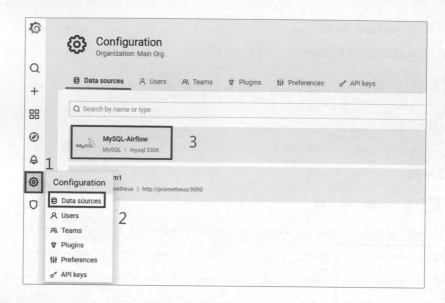

到這，成功架設 Airflow 與更新 Grafana 的 Data Source，是不是很簡單呢？都是前面章節照抄。完成以上步驟後，開始製作 Dashboard。

14.3.4.1　製作 Dashboard

原先想引用 Grafana 的 Dashboard 模板，但現成的模板太少，且沒有使用 MySQL 作為 Data Source，沒關係，筆者自己製作。

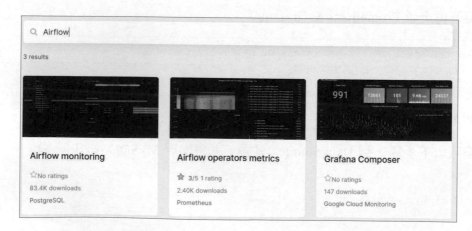

以下連結，就是筆者製作的 Airflow Dashboard（後續會介紹如何製作）。

https://github.com/FinMind/FinMindBook/blob/master/DataEngineering/
Chapter14/14.3/14.3.4/airflow.json

讀者可透過以上連結下載，並透過以下方式，導入 Dashboard。

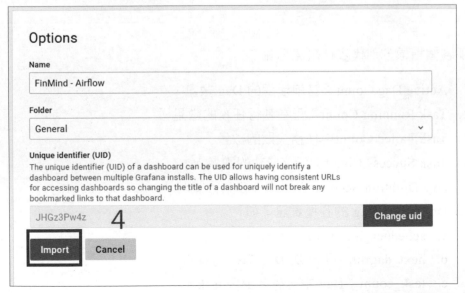

最後，就會出現下圖。

那各個指標分別代表什麼意思呢？

1. Active Dag Count：當前活躍的 Dag 數量。

2. Task Running Count：正在執行任務的數量。

3. Task Failed Count：任務失敗的數量。

4. Task Success Count：任務成功的數量。

5. Avg Duration Seconds：平均任務執行號時。

6. Dag Info：Dag 的各種資訊，如：

    a. scheduer_interval：排程設定。

    b. next_dagrun：下一次 Dag 執行的時間。

    c. task_count：Dag 中，有多少個 task。

7. Daily Dag Run Count：每天執行多少個 Dag。

8. Task Analysis：有一個漏斗，可篩選特定 Dag，主要用於分析，特定 Dag 下，每個任務的執行狀況，方便做效能分析。

以上是常見的監控指標，那要如何製作個人的 Dashboard 呢？我們接著往下看。

14.3.4.2 如何製作個人的 **Dashboard** ？

1. 首先，新增一個 Dashboard。

點選左邊菜單的 +，接著點選 Dashboard，之後會初始化一個 Dashboard，再點選 Add a new panel，製作第一個圖表。

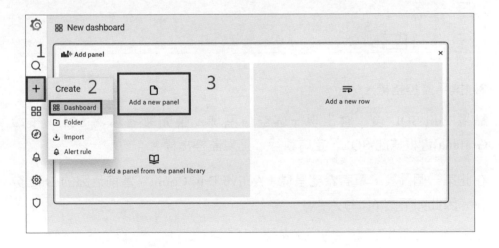

2. 設定資料源 Data Source。

由於前一節，新增了 MySQL 資料庫，因此在 Data source 的部分，選擇 MySQL-Airflow，接著點選 Edit SQL，來撰寫 SQL。

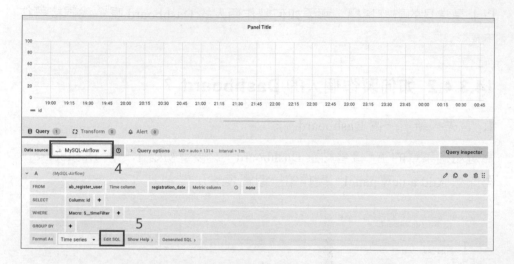

3. 撰寫監控指標。

點選 Edit SQL 後，會出現一個空白區塊，讓開發者填寫 SQL，後續 Grafana 會根據此 SQL，定時執行，呈現監控指標。

在此第一個圖表，筆者希望呈現，Active Dag Count - 當前活躍的 Dag 數量，因此如下圖 6，輸入：

```
SELECT count(1) FROM `dag` where is_active = 1
```

完成後，在 Format As 選擇 Table，代表我們要用 Table 呈現，最後在圖 8 選擇，Switch to table。

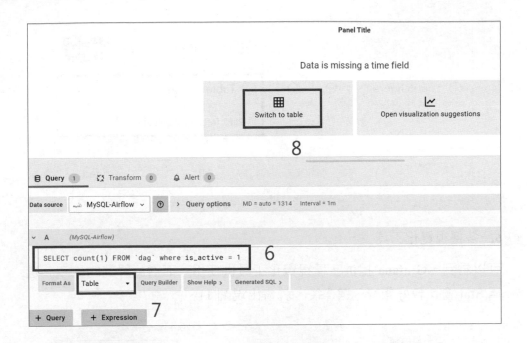

4. 最終 Grafana 根據以上提供的 SQL，將結果呈現在圖表上，如下，有 6 個活躍的 Dag。

Panel Title
count(1)
6

5. 客製化圖表顯示方式。

單純數字不好閱讀，所以在此點選右上角的 Table，選擇其他的視覺化方式。

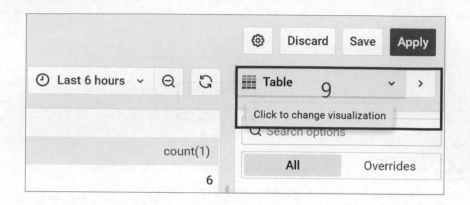

6. 多種視覺化工具。

可以看到，Grafana 提供多種視覺化工具供使用，而對於單純的數字，選擇 Stat 會比較好閱讀，選擇後，就會出現圖 11。

7. 設定標題。

最後在圖 12 的位置 - Title，輸入 Active Dag Count，方便判斷這指標的意義。

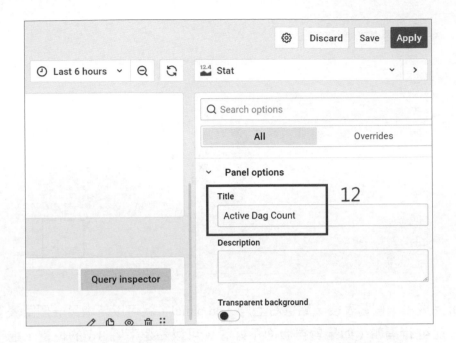

8. 完成後，在圖 13 的位置，就成功修改標題了。完成此圖後，點選左上角的箭頭，回到儀表板 Dashboard。

9. 可以看到，讀者做出第一個圖表了！如果覺得圖表太大，可以將滑鼠移到圖的右下角，進行縮放調整。

10. 完成第一個圖表後,點選右上角 Add panel,就可以新增另一個圖表,並進行編輯(如果覺得版面不好看,可以點選下圖 16 的位置,進行滑鼠拖曳移動)。

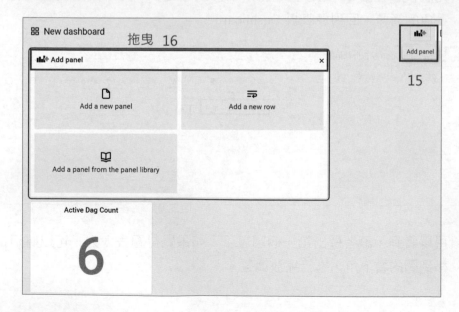

其餘的指標，讀者可回到前一節導入的儀表板，點選 Edit 查看背後的 SQL 語法，自行製作。

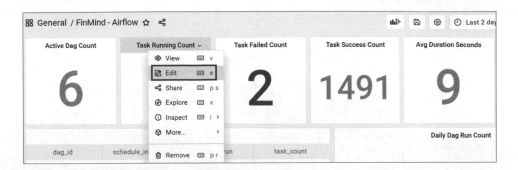

當以上指標製作完後，接著筆者開始製作。

❑ Dag Info 資訊

Dag Info			
dag_id	schedule_interval	next_dagrun	task_count
ABCD	null		
BashOperator	"*/5 * * * *"	2023-05-31 00:45:00	1
BranchPythonOperator	"*/5 * * * *"	2023-05-31 00:45:00	3
DockerOperator	null		
DummyOperator	"*/5 * * * *"	2023-05-31 00:45:00	6
HelloWorld	null		
PythonOperator	"*/5 * * * *"	2023-05-31 00:45:00	1
taiwan_stock_price	"0 17 * * *"	2023-05-30 17:00:00	4

‹ 1 ›   1 - 8 of 8 rows

步驟如下：

1. 先新增一個圖表 panel，並拖曳到下方的位置後，點選 Add a new panel。

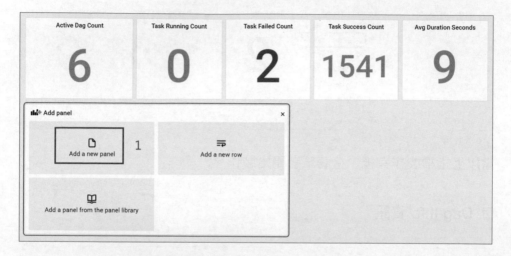

2. 在 Data source 選擇 MySQL-Airflow 後，一樣點擊 Edit。

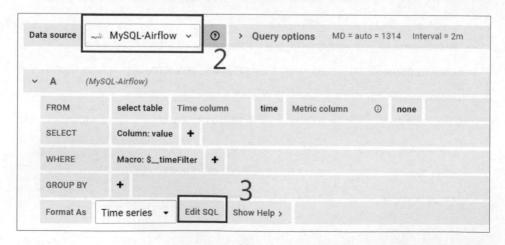

3. 輸入以下 SQL。

```sql
SELECT
    dag.dag_id,
    dag.schedule_interval,
    dag.next_dagrun,
    task_instance.task_count
FROM
    dag
LEFT JOIN(
    SELECT
        dag_id,
        COUNT(DISTINCT(task_id)) AS task_count
    FROM
        `task_instance`
    GROUP BY
        dag_id
) AS task_instance
ON
    dag.dag_id = task_instance.dag_id
```

最後在 Format As 選擇 Table。

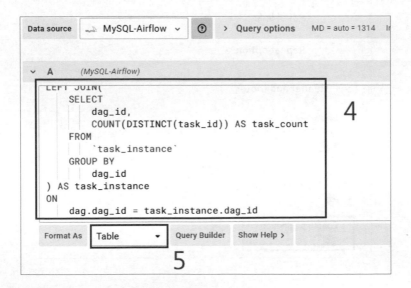

4. 接著在右上角，視覺化部分，選擇 Table，並在 Title 輸入 Dag Info，
最後在圖 8，點選 Enable pagination，啟用分頁功能。

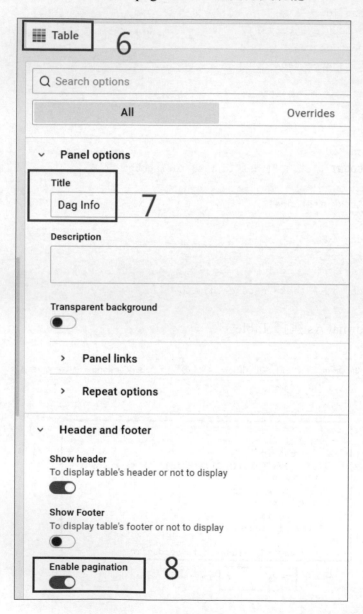

5. 調整欄位置中。

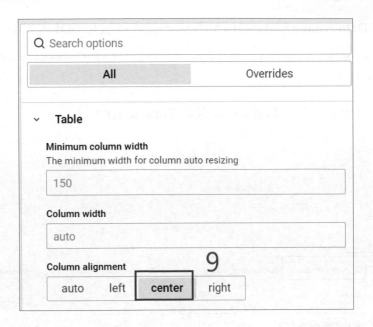

6. 到此，就完成 Dag Info 製作了！

Active Dag Count	Task Running Count	Task Failed Count	Task Success Count	Avg Duration Seconds
6	0	2	1591	8

Dag Info

dag_id	schedule_interval	next_dagrun	task_count
ABCD	null		
BashOperator	"*/5 * * * *"	2023-05-31 01:40:00	1
BranchPythonOperator	"*/5 * * * *"	2023-05-31 01:40:00	3
DockerOperator	null		
DummyOperator	"*/5 * * * *"	2023-05-31 01:40:00	6
HelloWorld	null		
PythonOperator	"*/5 * * * *"	2023-05-31 01:40:00	1

‹ 1 2 › 　　　1 - 7 of 8 rows

同樣邏輯，製作下列。

❑ Daily Dag Run Count

1. 設定 Data Source，MySQL-Airflow，輸入客製化 SQL，如下：

```
SELECT date(execution_date) as time, count(1) as count FROM `dag_run`
group by time
```

因為要看 Daily 資訊，Format As 選擇 Time series。

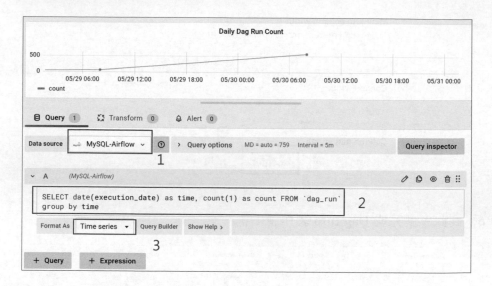

2. 接著視覺化一樣選 Time series，最後 Title 設定 Daily Dag Run Count。

3. 如果出現異常，原因是，因為 Daily 表是以天為單位的資料，因此須設定儀表版的時間區間，如下：

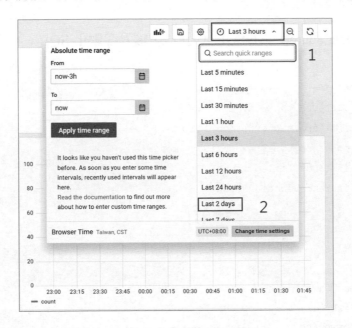

完成後如下圖。

（如果讀者 Airflow 剛開始運行，還不會有以下資料，讓 Airflow 執行超過 1 天，累積夠多任務執行數據，就會跟下圖一樣了。）

❏ 最後是 Task Analysis

也是一樣的方式。

1. 設定 Data Source，MySQL-Airflow，輸入客製化 SQL，如下：

```
SELECT
    dag.dag_id,
    task_instance.task_id,
    task_instance.avg_duration_seconds,
    task_instance.total_execution_times,
    task_instance.failed_count,
    task_instance.success_count
FROM
    dag
LEFT JOIN(
    SELECT
        dag_id,
        task_id,
        ROUND(AVG(duration)) AS avg_duration_seconds,
        COUNT(1) AS total_execution_times,
        SUM(IF(state = "failed", 1, 0)) AS failed_count,
      SUM(IF(state = "success", 1, 0)) AS success_count
    FROM
        `task_instance`
    GROUP BY
        dag_id,
        task_id
) AS task_instance
ON
    dag.dag_id = task_instance.dag_id
```

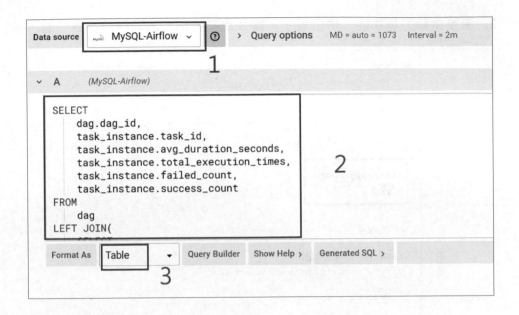

2. 選擇視覺化方式，並設定 Title，Task Analysis。

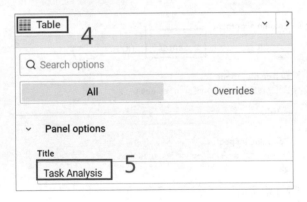

3. 最後啟動分頁功能，欄位置中，圖 8 比較特別，啟動欄位篩選功能。

最後就製作出，客製化的儀表板了！

最後進行儲存，點選右上角的 Save dashboard。

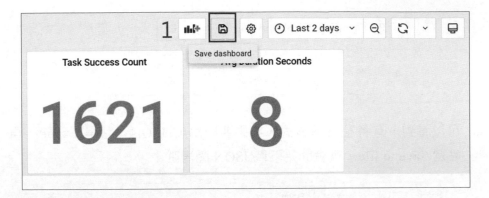

儀表板名稱，就由讀者自行決定了。

儲存後，要如何輸出 JSON 呢？

14.3.4.3 輸出 Dashboard 成 JSON

一般來說，都會輸出 JSON，最後上 git 做版控，那要如何輸出呢？

1. 點擊左上角的，分享圖示。

2. 可以看到，有各種分享方式，如果要輸出成 JSON，點選 Export，再點選 Save to file，就會開始下載 JSON 圖表啦！

是不是很簡單！現代化的 BI 視覺化工具，不論是 Redash、Superset、Grafana 等等，儀表板背後都是 JSON 架構，讀者未來即使使用本書以外的工具，也可以套用一樣的方式去做管理。

14.3.5 Traefik

最後，來到大魔王，也是與監控系統串接的最後一個服務，Traefik，還記得嗎？在 10.4，本書介紹使用 Traefik 進行反向代理，代理 API 的 DNS。

那為什麼要監控 Traefik 呢？還記得下圖嗎？由於使用 Traefik 後，所有的入口都經過 Traefik，因此 Traefik 會記錄所有服務的流量！這有助於幫助開發者，進行流量分析。

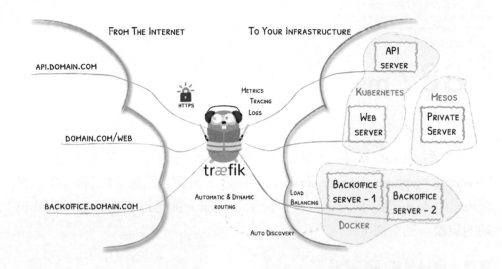

例如你架設了 API、BI 視覺化工具等多個服務，這時有一個統一的介面，協助分析流量，是非常助於管理系統的（這部份技能，還在 Data Engineer 的範疇內喔，Data Engineer 總是會需要架設 API、視覺化工具，本節就屬於後續的維運）。

下圖引用於 FinMind 的 Traefik Dashboard，或可參考 Grafana 的模板：

https://grafana.com/grafana/dashboards/12250-traefik-2-2/

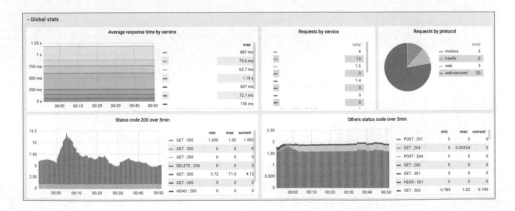

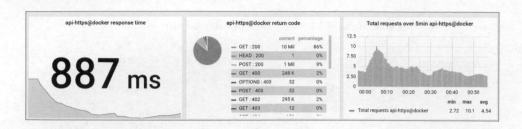

可以看到，Grafana 結合 Traefik 後，幫我們分析，平均回應耗時
（Average response time by service）、最近 5 分鐘的回應（Reponse）狀態
Status Code、最近 5 分鐘有多少流量等資訊。這可以協助評估，巔峰時
期的流量、回應速度。

最後，關鍵是如何架設呢？

這部份請參考以下連結：

https://github.com/FinMind/FinMindBook/tree/master/DataEngineering/
Chapter14/14.3/14.3.5

架構如下：

```
├──── Makefile
├──── README.md
├──── grafana
│     ├──── Dockerfile
│     ├──── dashboards
│     ├──── grafana-piechart-panel
│     │     ├──── LICENSE
│     │     ├──── MANIFEST.txt
│     │     ├──── README.md
│     │     ├──── dark.js
│     │     ├──── dark.js.map
│     │     ├──── editor.html
```

```
│   │       ├── img
│   │       │   ├── piechart-donut.png
│   │       │   ├── piechart-legend-on-graph.png
│   │       │   ├── piechart-legend-rhs.png
│   │       │   ├── piechart-legend-under.png
│   │       │   ├── piechart-options.png
│   │       │   ├── piechart_logo_large.png
│   │       │   ├── piechart_logo_large.svg
│   │       │   ├── piechart_logo_small.png
│   │       │   └── piechart_logo_small.svg
│   │       ├── light.js
│   │       ├── light.js.map
│   │       ├── module.html
│   │       ├── module.js
│   │       ├── module.js.LICENSE.txt
│   │       ├── module.js.map
│   │       ├── plugin.json
│   │       └── styles
│   │           ├── dark.css
│   │           └── light.css
│   ├── grafana.ini
│   └── provisioning
│       ├── dashboards
│       ├── datasources
│       │   └── all.yml
│       └── notifiers
├── grafana_prometheus.yml
├── prometheus
│   ├── Dockerfile
│   └── prometheus.yml
└── traefik.yml
```

因為與 14.3.4 高度重疊，在 14.3.5 就單純介紹 Traefik 的設置。基本上只需要調整：

1. prometheus.yml：讓 Prometheus 的 Targets 能看到 Traefik 的 metrics，並進行資料收集。
2. traefik.yml：這裡引用 10.4 的 yml 檔，https://github.com/FinMind/FinMindBook/blob/master/DataEngineering/Chapter10/10.4/traefik.yml，並增加一個設定，啟動 metrics 功能。
3. grafana-piechart-panel：由於 Grafana 的 Traefik 模板，需要使用 piechart 圖表，這裡本書也幫讀者準備好了，在建立 Grafana Image，自動將插鍵包進 Docker Image。

（本書在後面的章節，經常複製前面章節的程式，想告訴讀者的是，這是一步步演化的過程，當你想接觸新工具時，建議的方式是，在現有的架構下，去進行擴充，這樣才能將所學，完全融合在一起，蓋出一棟摩天大廈，而非 10 間只有一層樓的房屋。）

以上，只需三個步驟，就完成串接了。讀者可能會想問，要怎麼知道這些設定？本書監控系統部份到此，介紹了與 Netdata、MySQL、RabbitMQ 串接，基本上都是走 metrics 模式，現代化架構，都是類似的設計，而 Traefik 一定也有 metrics 功能，只需參考官方文件，即可得知如何啟動。

以下是 Traefik 的 yml 檔。

https://github.com/FinMind/FinMindBook/blob/master/DataEngineering/Chapter14/14.3/14.3.5/traefik.yml

```
version: "3.8"

services:
```

```
traefik:
  image: "traefik:v2.2" # image 版本
  command:
    # 設定顯示 dashboard
    - --api.insecure=true
    - --api.dashboard=true
    # log 模式，有 DEBUG、ERROR、INFO 等模式
    - --api.debug=true
    - --log.level=ERROR
    # https://doc.traefik.io/traefik/v2.2/routing/providers/docker/
    # 根據官方文件，有 docker、k8s 等模式可做選擇
    # 在此選擇 docker
    - --providers.docker=true
    - --providers.docker.endpoint=unix:///var/run/docker.sock
    - --providers.docker.swarmMode=true
    - --providers.docker.exposedbydefault=false
    - --providers.docker.network=traefik-public
    # 進入點，80、443 分別對應 http、https
    - --entrypoints.web.address=:80
    - --entrypoints.web-secured.address=:443
    # Grafana 監控系統的 metrics 位置
    - --metrics.prometheus=true
    - --metrics.prometheus.entryPoint=metrics
    - --entryPoints.metrics.address=:8082
    # SSL 設定
    - "--certificatesresolvers.myresolver.acme.httpchallenge=true"
    - "--certificatesresolvers.myresolver.acme.httpchallenge.
entrypoint=web"
    - "--certificatesresolvers.myresolver.acme.email=samlin266118
@gmail.com"
    - "--certificatesresolvers.myresolver.acme.storage=/letsencrypt/
acme.json"
```

```
ports:
  # 開對外 port，除了 80、443 是給 http、https 外
  # 8080 port 給 dashboard
  # 但我們 8080 已經給 phpmyadmin
  # 因此這裡 published 到 8888
  - target: 80
    published: 80
    mode: host
  - target: 443
    published: 443
    mode: host
  - target: 8080
    published: 8889
    mode: host
  # Grafana 監控系統的 metrics 位置
  - target: 8082
    published: 8082
    mode: host
volumes:
  # 存放 SSL 憑證
  - "letsencrypt:/letsencrypt"
  - "/var/run/docker.sock:/var/run/docker.sock:ro"
restart: unless-stopped
# log 最大 size
logging:
    driver: "json-file"
    options:
        max-size: "50m"
# deploy 設定
deploy:
  replicas: 1
  update_config:
```

```
      parallelism: 1
      delay: 10s
      order: stop-first
      failure_action: rollback
    placement:
      constraints: [node.role == manager]
  networks:
      - traefik-public
      - my_network

networks:
  traefik-public:
    external: true
  my_network:
    external: true

volumes:
  letsencrypt:
```

在第 25 行的位置，啟動了 prometheus 功能，且端點 port 是 8082。

https://github.com/FinMind/FinMindBook/blob/master/DataEngineering/
Chapter14/14.3/14.3.5/traefik.yml#L25

```
# Grafana 監控系統的 metrics 位置
- --metrics.prometheus=true
- --metrics.prometheus.entryPoint=metrics
- --entryPoints.metrics.address=:8082
```

設定端點在 8082 後，在 ports 的部分，也將 Docker 內部的 port 與外部串
接，如下：

https://github.com/FinMind/FinMindBook/blob/master/DataEngineering/
Chapter14/14.3/14.3.5/traefik.yml#L48

```
# Grafana 監控系統的 metrics 位置
- target: 8082
  published: 8082
  mode: host
```

最後，由於本節 Airflow 已使用 8888 port，做一個微調，將 Traefik 的
port 改成 8889。

https://github.com/FinMind/FinMindBook/blob/master/DataEngineering/
Chapter14/14.3/14.3.5/traefik.yml#L46

```
- target: 8080
  published: 8889
  mode: host
```

完成以上設定後，使用以下指令，部屬 Traefik，方法都與 10.4 一樣。

先建立 Network

```
docker network create --driver=overlay traefik-public
```

部屬

```
docker stack deploy -c traefik.yml tr
```

完成後，跳轉到 Traefik 的網頁：
http://localhost:8889/

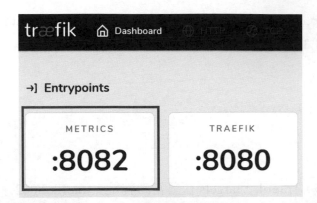

可以看到，出現一個 8082 的 METRICS，代表成功啟動 Prometheus Metrics 功能，接著跳轉到：

http://localhost:8082/metrics

```
# HELP go_gc_duration_seconds A summary of the GC invocation durations.
# TYPE go_gc_duration_seconds summary
go_gc_duration_seconds{quantile="0"} 3.83e-05
go_gc_duration_seconds{quantile="0.25"} 4.02e-05
go_gc_duration_seconds{quantile="0.5"} 4.89e-05
go_gc_duration_seconds{quantile="0.75"} 0.0003975
go_gc_duration_seconds{quantile="1"} 0.0033569
go_gc_duration_seconds_sum 0.0040588
go_gc_duration_seconds_count 7
# HELP go_goroutines Number of goroutines that currently exist.
# TYPE go_goroutines gauge
go_goroutines 55
# HELP go_info Information about the Go environment.
# TYPE go_info gauge
go_info{version="go1.14.8"} 1
# HELP go_memstats_alloc_bytes Number of bytes allocated and still in use.
# TYPE go_memstats_alloc_bytes gauge
go_memstats_alloc_bytes 4.619584e+06
# HELP go_memstats_alloc_bytes_total Total number of bytes allocated, even if freed.
# TYPE go_memstats_alloc_bytes_total counter
go_memstats_alloc_bytes_total 2.1071272e+07
# HELP go_memstats_buck_hash_sys_bytes Number of bytes used by the profiling bucket hash table.
# TYPE go_memstats_buck_hash_sys_bytes gauge
go_memstats_buck_hash_sys_bytes 1.457854e+06
# HELP go_memstats_frees_total Total number of frees.
# TYPE go_memstats_frees_total counter
```

是不是非常熟悉呢？這就是 Prometheus 習慣的格式。

下一步，在 prometheus.yml，新增 Targets 的設定檔，可參考以下連結。

https://github.com/FinMind/FinMindBook/blob/master/DataEngineering/
Chapter14/14.3/14.3.5/prometheus/prometheus.yml#L48

監控 Traefik

```
- job_name: Traefik
  static_configs:
  - targets: ['traefik:8082']
```

新增一個 job，Traefik，targets 位置當然就是，traefik:8082。

接著，按照以下步驟，更新 Prometheus。

```
docker stack deploy -c grafana_prometheus.yml monitor
```

跳轉到 Prometheus：
http://localhost:9090/targets

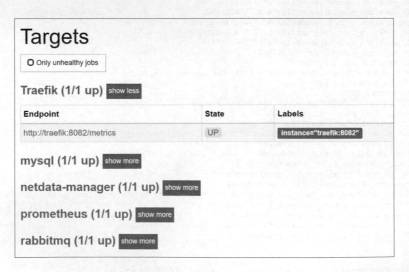

可以看到，Prometheus 成功收集到 Traefik 的 metrics。完成資料收集後，
開始製作 Dashboard。

14.3.5.1 製作 Dashboard

由於本書選用的 Traefik 圖表，需要額外 Grafana 的功能，因此在 Dockerfile 部份，做以下調整：

https://github.com/FinMind/FinMindBook/blob/master/DataEngineering/
Chapter14/14.3/14.3.5/grafana/Dockerfile

```
FROM grafana/grafana:8.5.25
COPY ./grafana/provisioning /etc/grafana/provisioning
COPY ./grafana/config.ini /etc/grafana/config.ini
COPY ./grafana/dashboards/* /var/lib/grafana/dashboards/
COPY ./grafana/grafana-piechart-panel /usr/share/grafana/plugins-bundled/
internal/piechart-panel
```

新增 piechart 圖表的插鍵。

完成插鍵升級後，直接引用 Traefik 的儀表板模板，https://grafana.com/
grafana/dashboards/12250-traefik-2-2/。

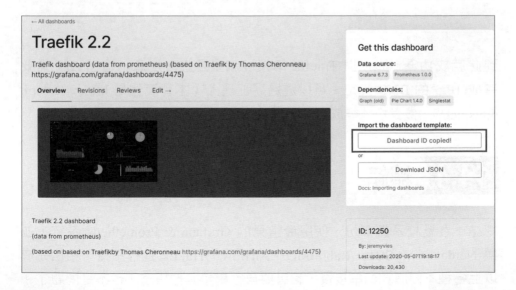

成功 Import 後，會出現以下畫面。

到此就成功囉！完成與 Traefik 串接、流量監控。注意，這裡 API 需要完成 10.5 的 DNS 設定，可以透過網址連上（本書範例是 https://testapi.ddns.net/），儀表板才能成功顯示喔。

14.4 總結

總結，在監控系統部份，使用最常見的 Grafana & Prometheus 系統，並與 Netdata、MySQL、RabbitMQ、Airflow、Traefik 進行串接，相信完成以上各種不同服務的串接後，讀者要自行串接其他服務，也不是問題了。

結論

本書以真實案例出發,一步步介紹所使用的工具,與對應的場景,不只是單純介紹工具,真實案例,有助於讀者理解,這些工具是為了解決什麼問題,之後面對其他真實問題,可以很輕易地做出架構圖,每個問題,讀者都有相對應的工具。

以本書來説,分成幾個部分:

1. 如何收集資料?

- 筆者初期做大數據分析,需要大量數據,為了解決 Data 問題,開始開發爬蟲,自動化收集資料。

- 在爬蟲過程中，遇到被 ban IP 問題，開始接觸分散式工具 RabbitMQ、Flower、Celery，分散到多台機器上，多個 IP，成功解決被 ban IP 問題。
- 分散式資料收集，每台機器各自收集資料，資料當然不可能存成 csv 等檔案，為了解決資料存放問題，開始架設 MySQL 資料庫，統一將爬蟲資料，存放到資料庫。

2. 如何提高效能？分散式架構、Docker 容器技術

- 分散式爬蟲使用了 RabbitMQ、Flower、MySQL 等多個服務，且分散式要部屬到多台機器上，過程複雜，因此使用 Docker 解決架設服務問題，且解決多機器部屬更新問題。
- 使用 Docker 後，為了解決管理多個 Docker 問題，開始接觸 Docker Swarm，並使用 Portainer 管理介面，統一管理多台機器。

3. 資料收集完了，如何提供資料？架設 API

- 資料收集完成後，要如何給其他人使用？這時開始接觸 API，讓任何語言都能輕易串接。
- 有 API 後，單純給對方 IP 讓對方發送 Request 嗎？太陽春，經過前面爬蟲的觀察後，多數網站的 API 都是使用網址，為了解決網址問題，開始接觸 No-IP，申請免費 IP。
- API 有網址後，卻缺少 SSL，導致網站安全性低，開始接觸免費 SSL 憑證 Let's Encrypt。
- 要讓 API 使用 SSL 憑證，多數使用 Nginx，但現代架構，跟容器技術結合，因此選用 Traefik，功能更齊全。
- 資料收集、API，都只停留在資料工程面，要如何應用呢？

4. 資料如何應用？視覺化

■ 開始架設視覺化工具 Redash，並製作 Dashboard，讓資料產品，跨出第一步，有視覺化分析工具，非常有助於資料分析產品化，不再只停留在工程面。

5. 多人開發的流程？ CICD

■ 當開發到此，需要增加開發能量，專案開始進入團隊、多人開發階段，自然而然地走進 CICD 的世界，為了解決，多人團隊開發上的問題，再每次的 PR 都做測試，並自動化部屬過程，讓團隊開發上，更加順暢、穩健。

6. 大數據領域，自動化、可視化的任務管理工具 - Airflow

■ 在大數據，特別是資料工程領域，大多都會使用排程管理工具，本書介紹的是最知名的工具 - Airflow，最後進階到，業界等級的，分散式 Airflow。

7. 架設了許多系統，如何維運？ Grafana & Prometheus 監控系統

■ 最後，當一個團隊、產品，架設各種服務後，一定會需要進行維運、監控，因此，本書也介紹了，最知名的、開源監控系統工具，並與本書介紹過的系統，都進行串接，讓開發者，可以在統一的頁面，監控多種服務的運作狀況，不用再跑到不同各種網頁上各別查看，這也是本書的核心之一，統一管理！

以上步驟，從資料收集開始，到分散式爬蟲、資料庫、API、Docker、CICD、Redash、Airflow、Grafana、Prometheus，成功從 0 到 1 打造出一

套資料分析產品。讀者可以根據自己有興趣的領域，做爬蟲收集資料，最後做出視覺化報表，完成專屬個人的分析作品，體驗產品開發的過程。

以下是幾個，讀者可以嘗試的分析作品例子。

例如：
- 對各大求職網站上，對職缺、薪資進行分析。
- 對各大電商平台，產品價格、產品種類分析。
- 對各大社群平台，分析群眾。

實際上，有非常多題目可以做，不用愁要選什麼題目做 Side Project，當你擁有以上技能後，要做什麼都不是問題。也可以跟筆者一樣，將資料做成 API、Open Data 方式，放到 GitHub 上，提供大眾使用。

如果對於本書有疑問，可以寄信到以下信箱，samlin266118@gmail.com，或是到 FB 搜尋 FinMind 社團，https://www.facebook.com/groups/401634838071226，筆者會一一回覆你。